Anatomy and Physiology
A Self-Instructional Course

Anatomy and Physiology
A Self-Instructional Course

1. The Human Body and The Reproductive System
2. The Endocrine Glands and The Nervous System
3. The Locomotor System and The Special Senses
4. The Respiratory System and The Cardiovascular System
5. The Urinary System and The Digestive System

1. The Human Body and The Reproductive System

Written and designed by
Cambridge Communication Limited

Medical adviser

Bryan Broom MB BS(Lond)
General Practitioner
Beit Memorial Research Fellow
Middlesex Hospital Medical Research School

SECOND EDITION

Churchill Livingstone
EDINBURGH LONDON MELBOURNE AND NEW YORK 1985

CHURCHILL LIVINGSTONE
Medical Division of Longman Group Limited

Distributed in the United States of America by Churchill Livingstone Inc., 650 Avenue of the Americas, New York, N.Y. 10011, and by associated companies, branches and representatives throughout the world.

First edition 1977
Second edition 1985
Reprinted 1991, 1993

ISBN 0-443-03170-3

British Library Cataloguing in Publication Data
A catalogue record for this book is available from the British Library

Library of Congress Cataloging in Publication Data
Anatomy and physiology.
Rev. ed. of: anatomy and physiology / Ralph Rickards, David F. Chapman. 1977.
Contents: 1. The human body and the reproductive system — 2. The endocrine glands and the nervous system — 3. The locomotor system and the special senses — [etc.]
1. Human physiology — Programmed instruction. 2. Anatomy, Human — Programmed instruction. I. Broom, Bryan. II. Rickards, Ralph. Anatomy and physiology. III. Cambridge Communication Limited.
QP34.5.A47 1984 612 84-4977

The publisher's policy is to use **paper manufactured from sustainable forests**

Printed in Hong Kong
WC/03

Contents

Contents

The Human Body

1. Introduction

1.1. The cell

All living matter is made up of *cells,* the smallest units of life which can function and reproduce themselves.

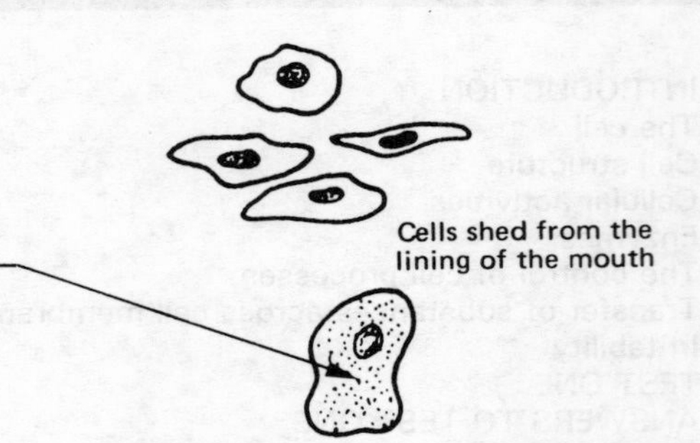

The smallest animals, such as the pond-dwelling **amoeba**, consist of a single cell which is capable of dealing with the varying temperature, chemical composition etc., of its environment and of performing all the necessary functions of animal life.

Man consists of sixty thousand billion (6×10^{16}) cells. These cells share the various essential functions among themselves by becoming specialised. They have become dependent on an environment around them which is closely controlled.

Mast cells are about 10–30 μm in diameter (1000 μm = 1mm). White blood cells are 10–12 μm across, the cell body of the larger nerve cells of the brain 50 μm and the ovum (the egg cell from which an individual develops) is 150 μm across.

All cells have a similar basic structure. They are bounded by a thin, but complex, **cell membrane** which carefully controls the entry and release of substances from the cell. The substance of the cell is called **protoplasm.**

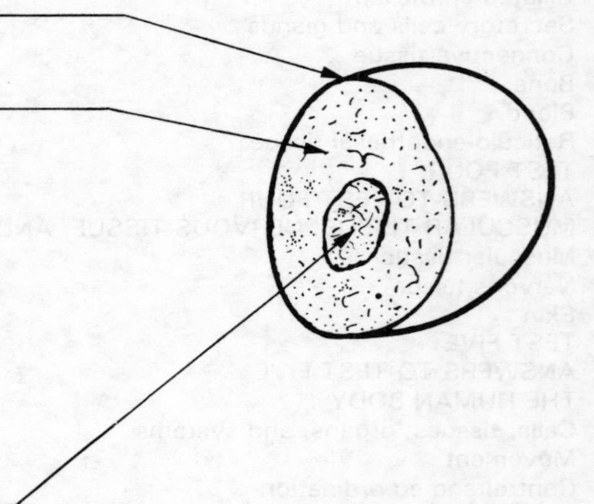

Protoplasm is a highly complex material. It has a framework of large *protein* molecules in which *lipids* and *carbohydrates* are included. These exist in a dilute solution of inorganic ions, or electrolytes. The principal metallic ion (positively charged cation) is potassium. Chloride, bicarbonate and phosphate are the principal negatively charged ions (anions).

Every cell capable of division has a controlling centre, the **nucleus.** The nucleus contains a 'blueprint' of all the activities of the cell (in the form of long molecules of the nucleoprotein, deoxyribonucleic acid, DNA).

The cells lie among supporting tissue, the *intracellular substances*, such as fibrous tissue. All the spaces between cells are filled with fluid. The composition of this *extracellular fluid* is controlled within narrow limits. The principal cation of the extracellular fluid is sodium.

1.2. Cell structure

The *cytoplasm,* the protoplasm outside the nucleus, contains many small, complex structures which can only be seen with an electron microscope. The function of many of these is poorly understood.

The cell membrane is a molecular sandwich of protein molecules (/\/\ ) with a double-layered arrangement of phospholipid (ό ό) forming the filling. The phospholipids have a water soluble end (o) and a fat soluble end (l) and these are arranged so that the only large molecules which can enter the cell directly are fat soluble molecules.

The **endoplasmic reticulum** is a double-layered arrangement of membrane like a flattened empty bag. It is found throughout the cell, dividing it up into functional areas.

The **Golgi apparatus** is a laminated arrangement of membranes close to the nucleus. It appears to be important for secretion.

The **centrosomes**, a pair of crossed cylinders, are essential for cell division.

The **ribosomes** are the site of production of all the proteins. They are largely composed of ribonucleic acid (RNA). They lie free or on the endoplasmic reticulum.

The **lysosomes** are essentially packets of enzymes, which can 'digest' droplets of material within the cell.

Secretory or **storage droplets**, often within a membrane (vesicles), are seen within the cytoplasm. They are composed of lipid, carbohydrate or protein.

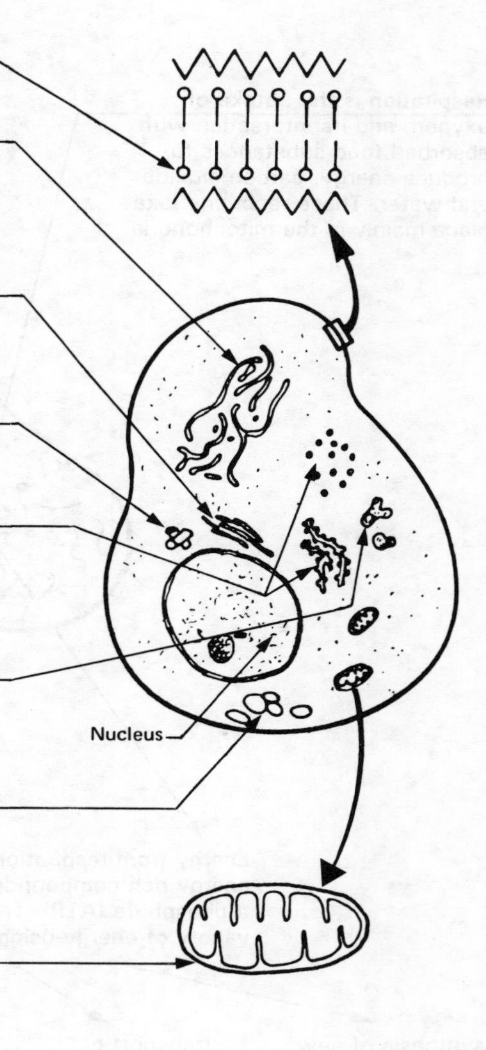

The **mitochondria** are double-layered tubes with internal partitions on which are packed arrays of enzymes. These enzymes are capable of converting simple foods into energy-rich compounds. They are the 'power houses' of the cell.

The **nuclear membrane** is a double-layered coat around the nuclear material. It is really an extension of the endoplasmic reticulum with thin pores in it.

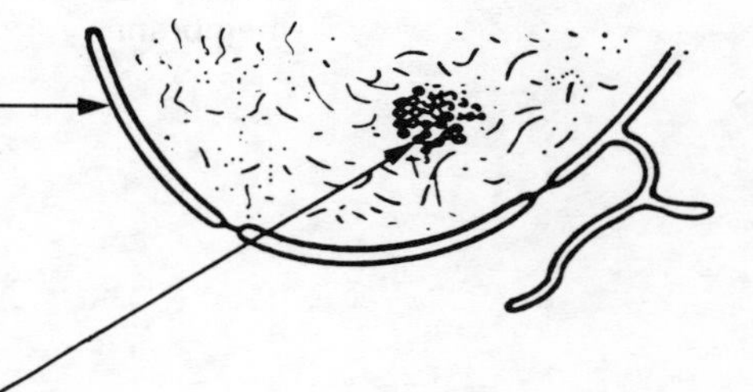

The nuclear material is simple in overall structure. It consists of a few immensely long DNA molecules packed randomly together like a bundle of string. Within it is often seen a body composed of RNA, the **nucleolus**.

1.3. Cellular activities

The activities of the cell are classified as absorption, excretion, irritability, conductivity, contractility, growth and reproduction.

Metabolism is the general term used to describe all the physical and chemical processes which occur in protoplasm and which support life. Metabolic processes are either *anabolic*, utilising energy to build up new substances, or *catabolic*, breaking down substances to release energy.

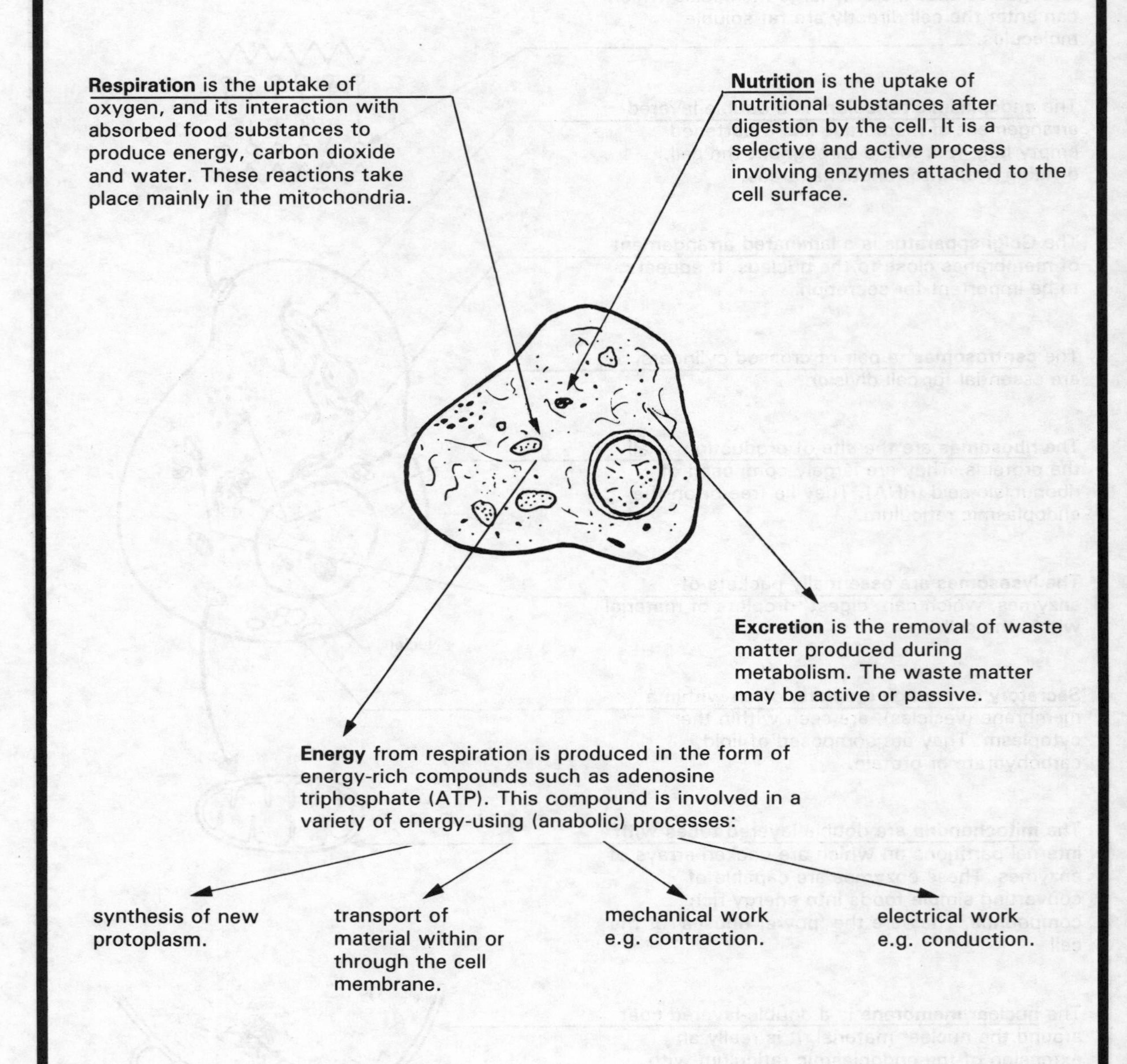

1.4. Enzymes

The activities of the cell are governed, to a large extent, by *enzymes*.

Enzymes are catalysts which permit chemical reactions to occur quickly within the cell. Without enzymes the reactions would only occur very slowly.

All enzymes (of which protoplasm is largely composed) are complex proteins. Their large molecules are so shaped as to bring reacting substances closely together on the enzyme surface, so speeding their reaction:

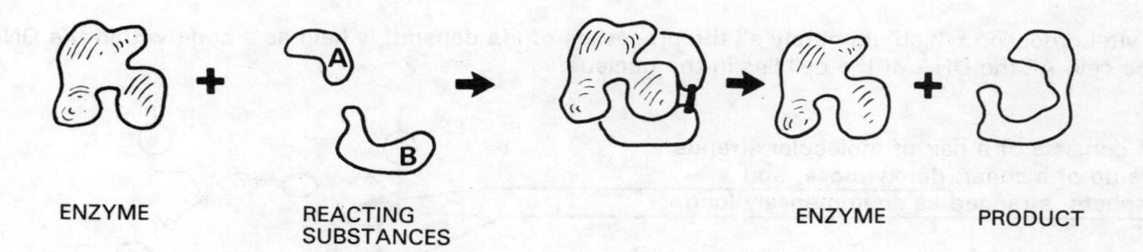

Enzymes are very specific; they usually control one reaction only. They have an ideal pH and temperature at which they are most active. Extremes of pH or temperature destroy enzymes by distorting their molecular shape.

Enzymes are able to operate outside cells in artificial conditions.

1.5. The control of cell processes

The nature and rate of metabolic processes within the cell are controlled by enzymes. The production of enzymes is in turn controlled by the nucleus.

Enzymes are complex proteins — chains of amino acids. There are over 20 different amino acids each with a fairly simple structure. A typical enzyme molecule contains hundreds or thousands of amino acids linked together and this chain is folded up on itself to produce a bundle with a specific shape.

This outward shape depends simply on the precise order of its chain of amino acids.

The vital order, on which ultimately all the processes of life depend, is held as a code within the DNA of the cell. All the DNA of the cell lies in the nucleus.

DNA consists of a pair of molecular strands made up of a sugar, **deoxyribose**, and **phosphate**, arranged as an immensely long chain.

Joining the two strands, like the rungs of a ladder, is a pair of **bases**.

There are four bases in DNA, adenine (A), thymine (T), guanine (G) and cytosine (C). Because of the nature of their bonds, adenine is always paired with thymine, and cytosine with guanine.

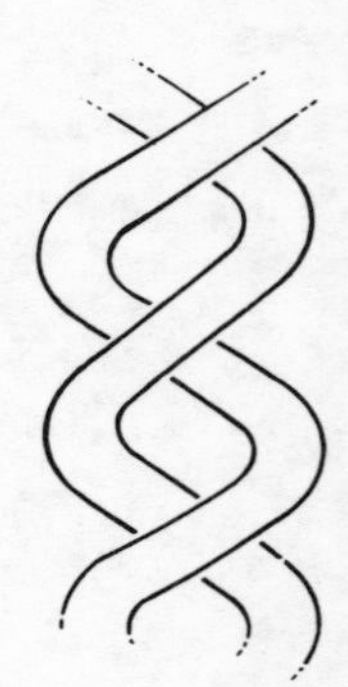

The sugar-phosphate strands spiral round the central bases like the coils of a spring — a double helix.

The sequence of bases on the DNA molecule is the code for the order of amino acids in the particular protein made by the cell.

Each sequence of three bases *(triplet)* is the code for a single amino acid. For example the amino acid *valine* is represented by the code GTA.

There are $4^3 = 64$ possible triplets so some amino acids have several codes. There are also codes to represent the beginning and end of a protein chain.

If the DNA molecule splits into two halves by opening between the bases, the single strands formed are rather like mirror images of each other. These strands provide a template from which *two* molecules, identical to the parent molecule, can be built from fresh deoxyribose, phosphate, bases, and from the original strands.

This is how new cells obtain precise 'genetic information' at each division.

An opened DNA molecule can also be used as a template for the construction of a relatively short molecular stand of only a few thousand bases. This single-stranded uncoiled molecule can act as a mobile 'site plan' outside the nucleus. It is known as *messenger RNA* (ribonucleic acid).

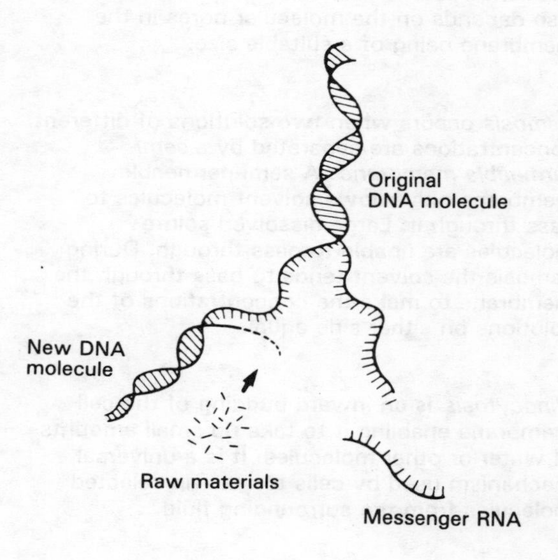

The messenger RNA passes into the cytoplasm where it runs over the surface of the ribosomes. On the ribosomes free amino acids are joined together, according to the base sequence, to form a protein chain.

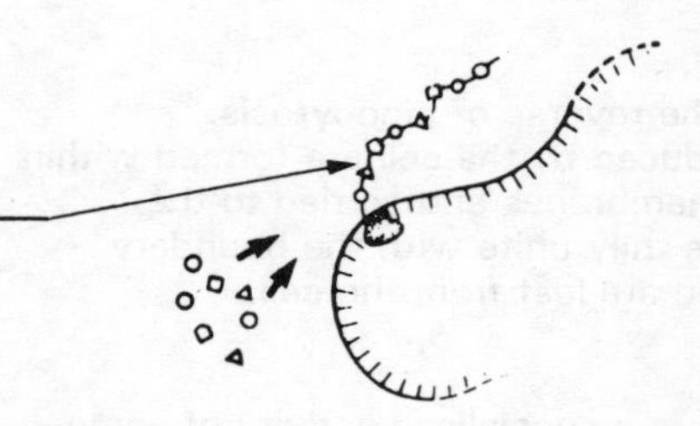

ATP is used up in this *anabolic* process.

Proteins for the use of the cell itself in the formation of protoplasm are formed on ribosomes lying free in the cytoplasm. The proteins destined for secretion are formed on ribosomes attached to endoplasmic reticulum, and are secreted into the cavity of the reticulum. These packets of protein, *secretory vesicles*, are thus immediately segregated from the cell.

1.6. Transfer of substances across cell membranes

A living cell is in a state of *dynamic equilibrium*. Although the components of the protoplasm remain relatively constant, substances are continuously entering and leaving the cell by a variety of processes.

Filtration occurs through permeable membranes when there is a pressure difference between the fluids on either side. Fluid and any molecules small enough to pass through the pores in the membrane are forced through. For example, the blood pressure in the capillaries causes substances to be forced through the lining cells into the tissues. Water and ions pass through, but amino acids do not.

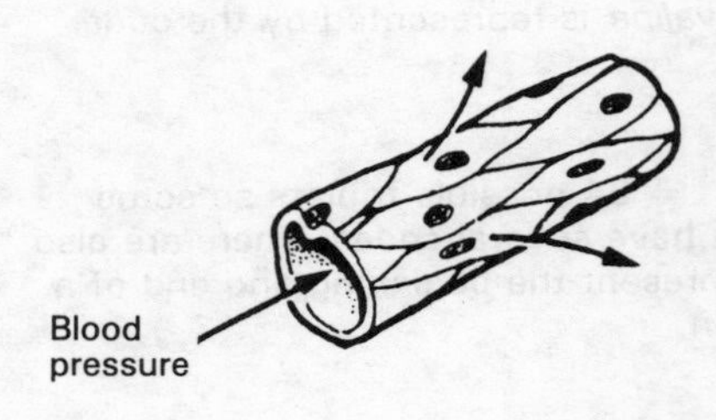

Diffusion is the passive movement of a substance from an area where it is highly concentrated to an area where it is less highly concentrated. For example, oxygen diffuses from the air spaces of the lung (high concentration) into the blood cells in the lung capillaries (low concentration). This movement also depends on the molecular pores in the membrane being of a suitable size.

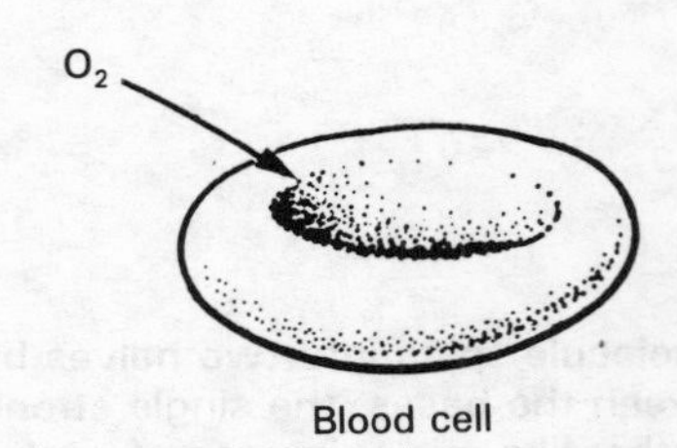

Osmosis occurs when two solutions of different concentrations are separated by a *semi-permeable membrane*. A semi-permeable membrane only allows solvent molecules to pass through it. Large dissolved solute molecules are unable to pass through. During osmosis the solvent tends to pass through the membrane to make the concentrations of the solutions on either side equal.

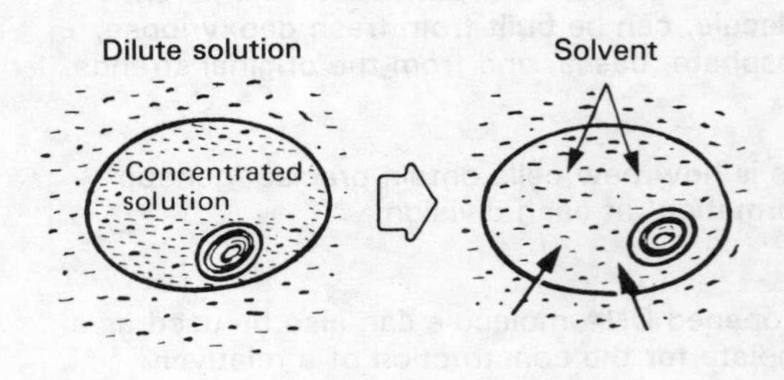

Pinocytosis is an inward budding of the cell membrane enabling it to take up small amounts of water or other molecules. It is a universal mechanism used by cells to take up selected molecules from the surrounding fluid.

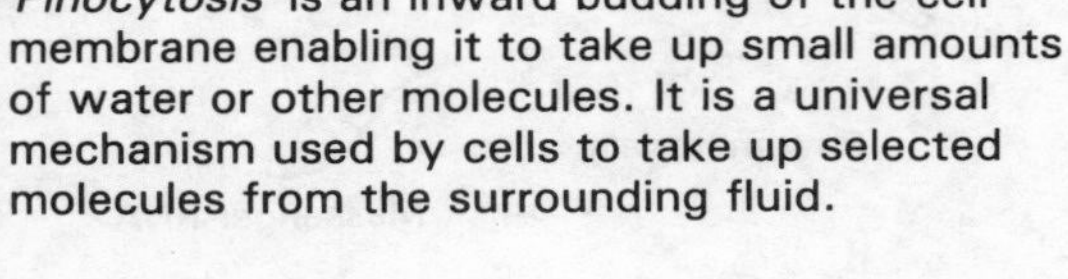

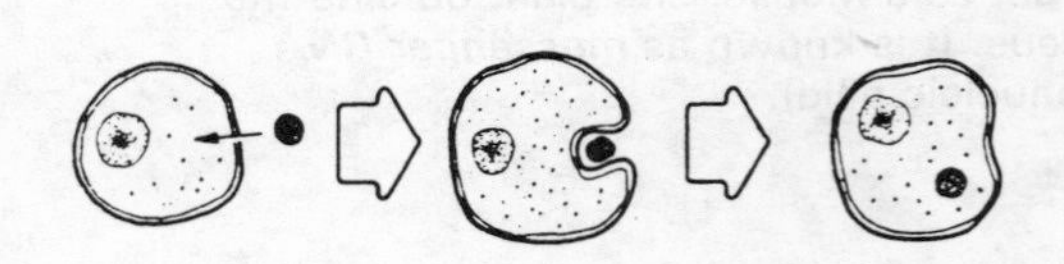

Secretion is the reverse of pinocytosis. Materials produced by the cell are formed within intracellular membranes and carried to the surface where they unite with the boundary membrane and are lost from the cell.

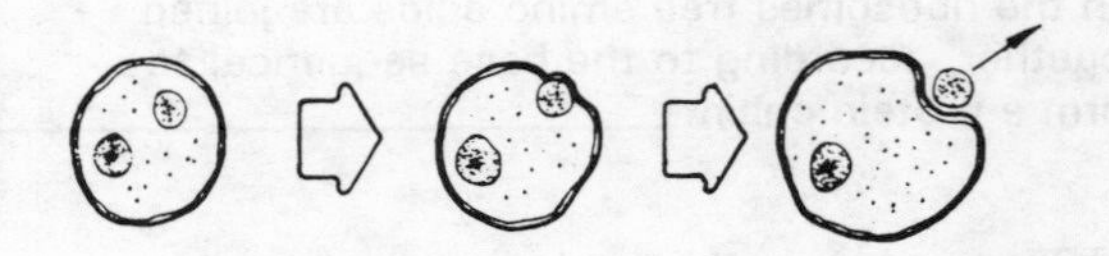

Phagocytosis is a specialised activity of some cells, such as white blood cells, which engulf and later ingest microorganisms, other cells and foreign particles.

1.7. Irritability

Irritability is the property which cells possess of responding to changes in environment. The degree and nature of the response depends on the specilaised functions of the cell.

Many simple unspecialised cells are capable of moving through fluids and tissue spaces in the body.

Movement is due to the formation of **pseudopodia** by the cell membrane. Cytoplasm streams into the pseudopodium with the result that the cell travels in the direction of the formation of the pseudopodium.

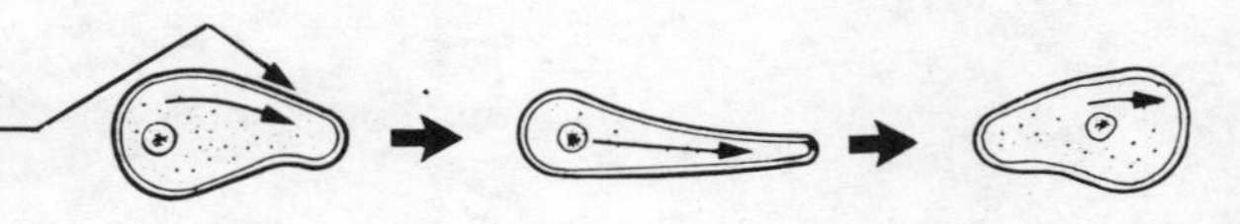

This type of movement is called *amoeboid*, as it is characteristic of the amoeba. White blood cells can move in this way.

Contractility
Some cells, although fixed in place and incapable of independent movement, can *contract* in response to stimuli. Contraction is due to a complex sequence of events inside the cell which are manifested either by the cell shortening, or by an increase in tension in the cell wall. It is characteristic of muscle cells.

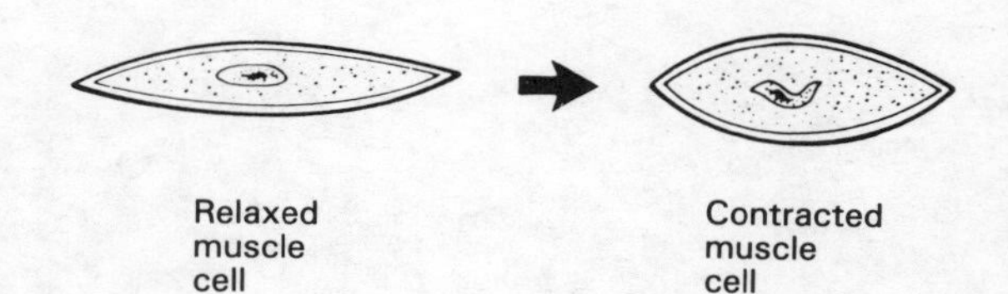

Conductivity
All cells have an electrical charge on their surface due to a constant expulsion of sodium ions from the cell by the cell membrane. Changes in the environment, or trivial damage can cause a wave of alteration of this charge to travel over the cell surface.

Nerve cells are specialised to conduct such electrical activity along long processes. Messages are carried rapidly from one part of the body to another by these cells.

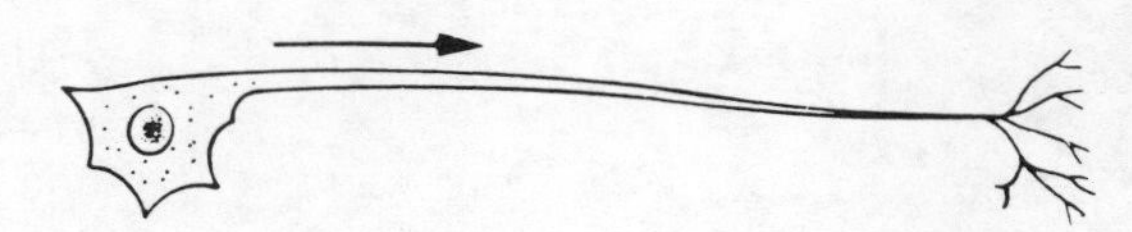

TEST ONE

1. Complete the following statements.

(a) Mast cells are about __________ in diameter.
(b) __________ are the site of protein production within the cell.
(c) Enzymes are the substances of which __________ is largely composed.
(d) DNA molecules constitute a 'blue-print' of __________ of a cell.

2. Indicate which of the names in the list below refer to the parts of the cell labelled on the diagram alongside, by placing the appropriate letters in the brackets.

1. Nucleolus. ( )
2. Golgi apparatus. ( )
3. Mitochondria. ( )
4. Endoplasmic reticulum. ( )

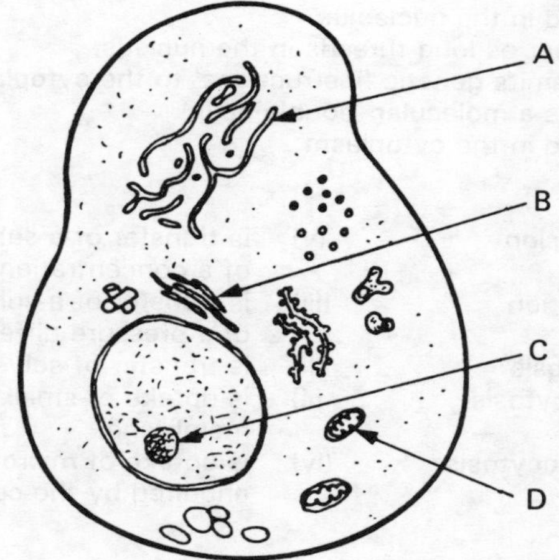

3. Indicate which of the following are characteristics of DNA and which are characteristics of RNA by placing ticks in the appropriate brackets.

		DNA	RNA
(a)	Found in the nucleolus.	( )	( )
(b)	Present as long threads in the nucleus.	( )	( )
(c)	Transmits genetic 'instructions' to the cytoplasm.	( )	( )
(d)	Forms a molecular 'double helix'.	( )	( )
(e)	Found in the cytoplasm.	( )	( )

4. Which of the statements on the right apply to the processes listed on the left?

(a) Diffusion
(b) Filtration
(c) Osmosis
(d) Pinocytosis
(e) Phagocytosis

(i) is transfer of solvent across a semi-permeable membrane.
(ii) is uptake of small particles by inward budding of the cell membrane.
(iii) is transfer of a substance across permeable membranes because of a pressure difference.
(iv) is uptake of microorganisms, other cells and foreign particles engulfed by the cell.
(v) is transfer of a substance across permeable membranes because of a concentration difference.

ANSWERS TO TEST ONE

1. (a) Mast cells are about *10–30 μm* in diameter.
(b) *Ribosomes* are the site of protein production within the cell.
(c) Enzymes are the substances of which *protoplasm* is largely composed.
(d) DNA molecules constitute a 'blue-print' of *all the activities* of a cell.

2.

1. Nucleolus.	(C)
2. Golgi apparatus.	(B)
3. Mitochondria.	(D)
4. Endoplasmic reticulum.	(A)

3.

		DNA	RNA
(a)	Found in the nucleolus.	( )	(√)
(b)	Present as long threads in the nucleus.	(√)	( )
(c)	Transmits genetic 'instructions' to the cytoplasm.	( )	(√)
(d)	Forms a molecular 'double helix'.	(√)	( )
(e)	Found in the cytoplasm.	( )	(√)

4.

(a)	Diffusion	(v)	is transfer of a substance across permeable membranes because of a concentration difference.
(b)	Filtration	(iii)	is transfer of a substance across permeable membranes because of a pressure difference.
(c)	Osmosis	(i)	is transfer of solvent across a semi-permeable membrane.
(d)	Pinocytosis	(ii)	is uptake of small particles by inward budding of the cell membrane.
(e)	Phagocytosis	(iv)	is uptake of microorganisms, other cells and foreign particles engulfed by the cell.

2. Cell reproduction

2.1. Mitosis

Cells do not continue to grow indefinitely, but divide into two cells when they reach their maximum size. *Mitosis* is the name given to the mechanism of ordinary cell division. It ensures that DNA is divided equally between the two new cells.

The long DNA molecules within the nucleus contract down by coiling on themselves during division to produce visible bodies called *chromosomes*. Since the DNA 'copies itself' during growth the chromosomes are able to split into identical *chromatids*.

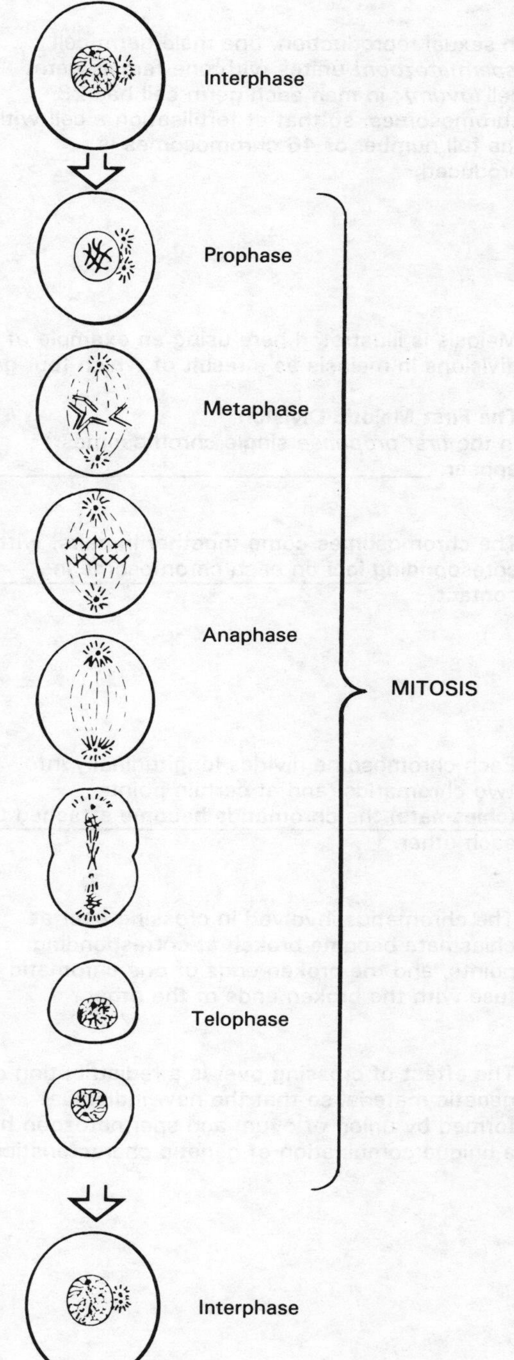

Prophase is the first stage of cell division. The nuclear chromatin forms distinguishable chromosomes. The nuclear membrane breaks down. The centresomes migrate to opposite poles of the cell.

At *metaphase*, the chromosomes arrange themselves around the equator of the cell.

At *anaphase* each chromosome divides longitudinally into two chromatids. The chromatids migrate to opposite poles of the cell.

Each set of chromatids becomes the chromosomes of the two new cells that will form.

At *telophase* a division appears across the middle of the cell and extends until the original cell splits into two smaller cells.

The chromosomes change back to chromatin threads, and the nuclear membrane and nucleolus reappear.

The period of cellular growth and activity between divisions is called *interphase*. During this period the DNA is uncoiled and not visible by microscopy as chromosomes. It is duplicating itself, so that at the next division a pair of chromatids is again present.

2.2. Meiosis

Examination of the dividing human cell shows that there are always 46 chromosomes, which by their division in mitosis pass 46 chromatids to each daughter cell. From their shape it is possible to arrange these 46 chromosomes into 23 pairs. 22 of these pairs are identical. These are the *autosomes*. The remaining pair is identical in females (a pair of X chromosomes), but differs in the male (one X and one Y chromosome). The X and Y chromosomes are the sex chromosomes. Possession of a Y chromosome in the body cells confers masculinity.

Every normal body cell contains a double (*diploid*) set of chromosomes which are derived from maternal and paternal sources. *Meiosis* is the mechanism whereby the number of chromosomes is halved in the production of germ cells, or reproductive cells.

In sexual reproduction, one male germ cell *(spermatozoon)* unites with one female germ cell *(ovum)*; in man each germ cell has 23 chromosomes, so that at fertilisation a cell with the full number of 46 chromosomes is produced.

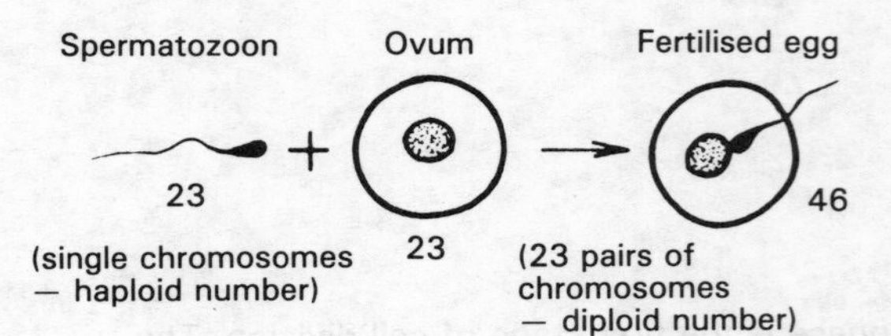

Meiosis is illustrated here using an example of a cell with one pair of chromosomes. There are two cell divisions in meiosis as a result of which four gametes are produced from one original cell.

The First Meiotic Division
In the *first prophase* single chromosomes appear.

The chromosomes come together in pairs, with coresponding **loci** on each chromosome in contact.

Each chromosome divides longitudinally into two chromatids, and at certain points **(chiasmata)** the chromatids become attached to each other.

The chromatids involved in crossing over at chiasmata become broken at corresponding points, and the broken ends of one chromatid fuse with the broken ends of the other.

The effect of crossing over is a redistribution of genetic material so that the new individual formed by union of ovum and spermatozoon has a unique combination of genetic characteristics.

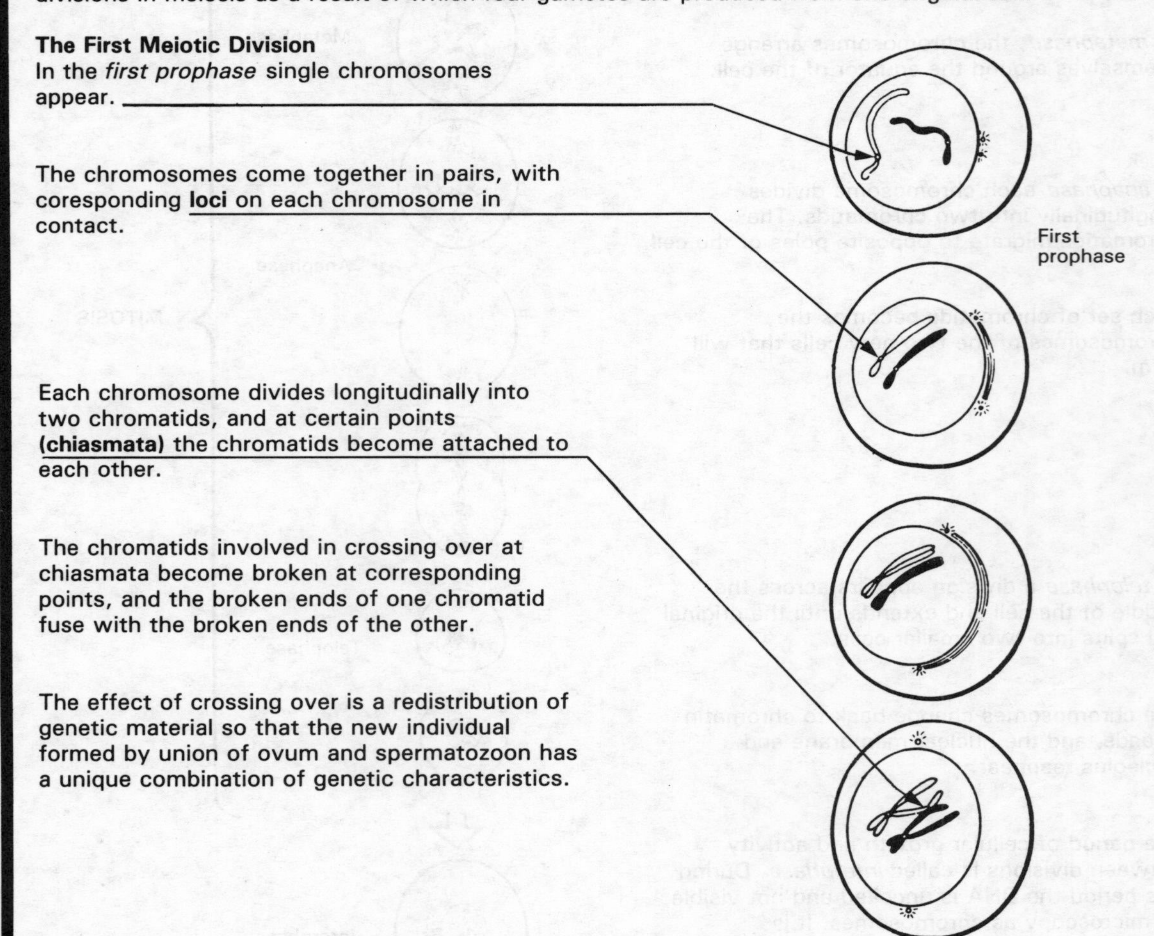

First metaphase
The chromatids arrange themselves around the equator of the cell.

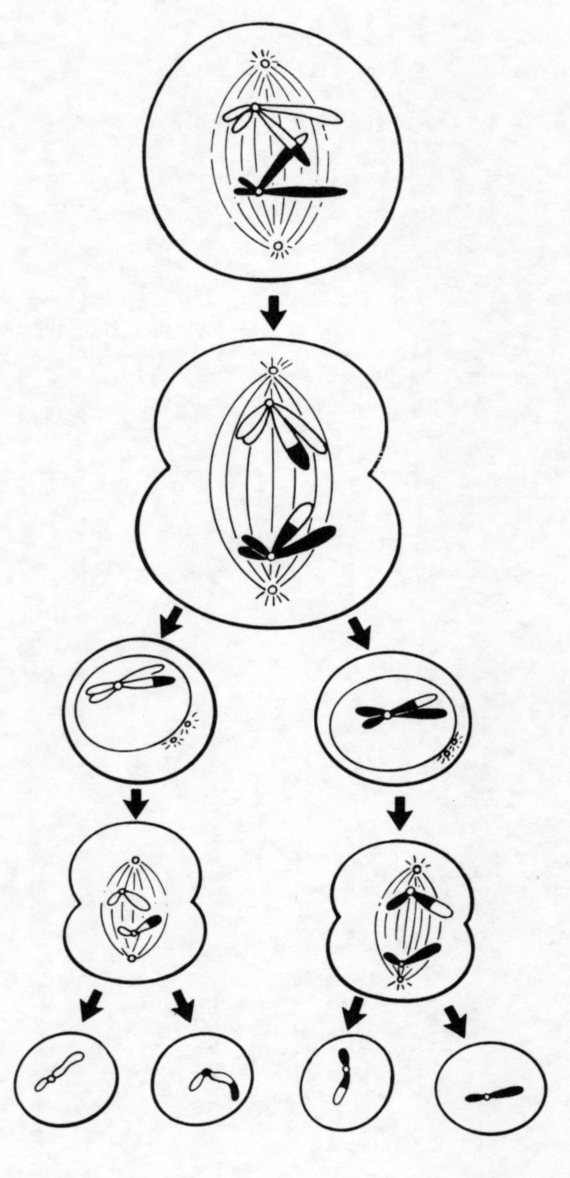

First anaphase
The chromatids separate and migrate to opposite ends of the cell.

First telophase
The two new cells separate.

The Second Meiotic Division
Each cell divides again (going through a second metaphase, anaphase, and telophase).

The end result in the male is four *gametes* which develop into four *spermatozoa* (see p. 58).

In the female three of the 'cells' formed are, in fact, just tiny masses of nuclear material which are discarded. All the original cytoplasm accompanies one gamete and forms a single large *ovum*. All the germ cells in the female have developed in her ovaries before birth and reached the stage of crossing over. Further development then stops for many years until at puberty the cell is released from the ovary. Meiosis then rapidly proceeds again and is completed at ovulation when the mature ovum may or may not be fertilised by a spermatozoon.

TEST TWO

1. Are the following statements true or false?

		True	False
(a)	DNA duplicates itself during the process of mitosis.	( )	( )
(b)	At metaphase the chromosomes arrange themselves around the equator of the cell.	( )	( )
(c)	Telophase is the final stage of mitosis.	( )	( )
(d)	The two cells resulting from mitosis contain identical DNA.	( )	( )

2. Which of the statements on the right apply to the items listed on the left?

(a) The Y chromosomes.
(b) The chiasmata.
(c) The diploid number.

(i) is the number of chromosomes found in all ordinary body cells.
(ii) confer masculinity.
(iii) link chromatids.

3. Indicate which of the names in the list below refers to the parts labelled on the diagram alongside, by placing the appropriate letters in the brackets.

1. Chiasmata. ( )
2. Centrosome. ( )
3. Chromatid. ( )

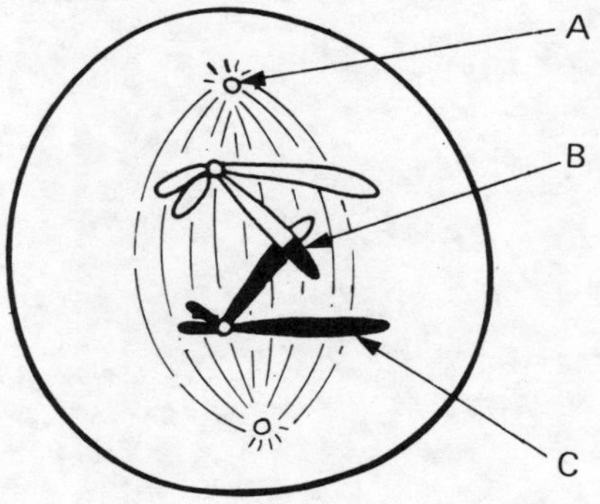

4. When is the chromosome number halved in meiosis?

(a) During the first meiotic division.
(b) During the second meiotic division.

ANSWERS TO TEST TWO

1.

		True	False
(a)	DNA duplicates itself during the process of mitosis.	( )	(√)
(b)	At metaphase the chromosomes arrange themselves around the equator of the cell.	(√)	( )
(c)	Telophase is the final stage of mitosis.	(√)	( )
(d)	The two cells resulting from mitosis contain identical DNA.	(√)	( )

2. (a) The Y chromosomes — (ii) confer masculinity.
(b) The chiasmata — (iii) link chromatids.
(c) The diploid number — (i) is the number of chromosomes found in all ordinary body cells.

3. 1. Chiasmata. (B)
2. Centrosome. (A)
3. Chromatid. (C)

4. (a) During the first meiotic division.

3. Maturation, differentiation and organisation of cells

3.1. Maturation and differentiation

The specialised functions of a cell determine its shape and structure. During the development of a fertilised egg into an embryo by repeated cell divisions cell *differentiation* occurs, during which many different types of cell develop.

Once full differentiation has occurred, the cell loses the ability to change its nature at further divisions.

Some cells have only a short life, for example, red blood cells. These cells are replaced as they die. Other cells, for example nerve cells, do not divide and remain fixed once development is complete.

Some examples of different cells, drawn to scale, are shown below:

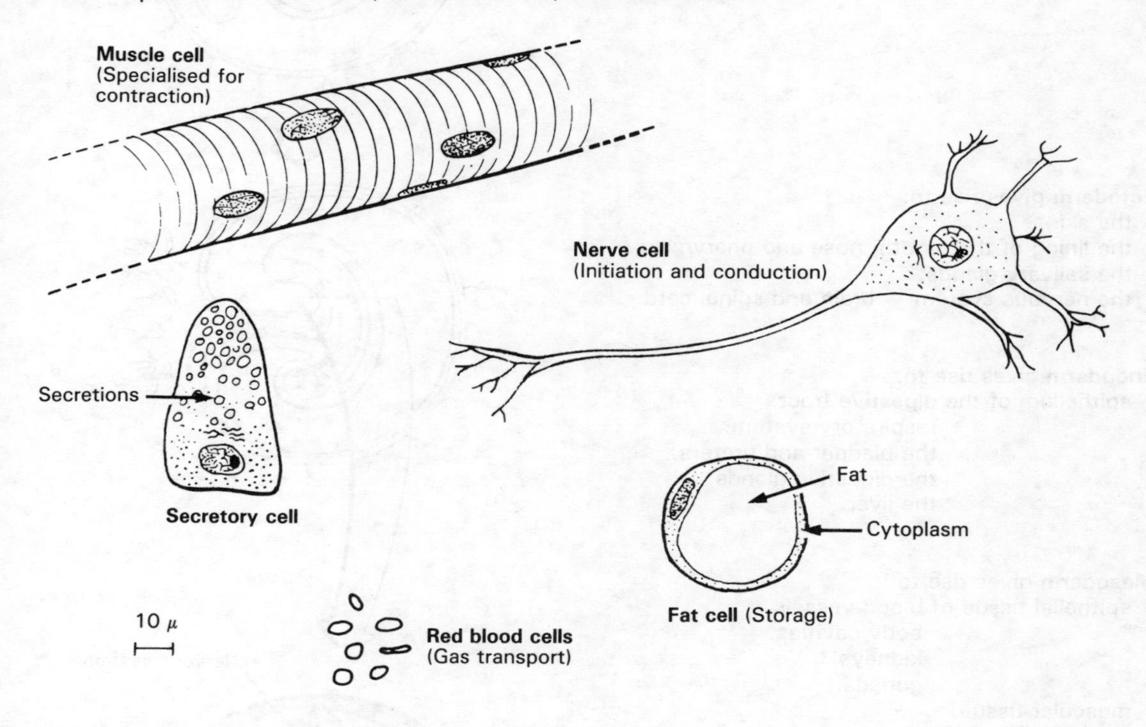

Cells which have the same function are united together in *tissues*:

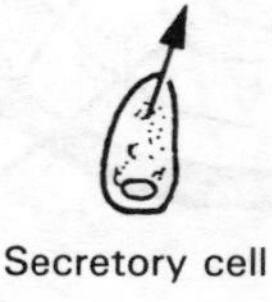

Secretory cell

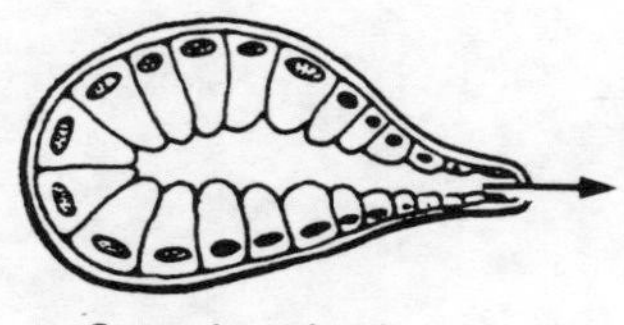

Secreting gland

3.2. The organisation of the tissues

The embryo develops from a plate of tissue which lies between two fluid-filled sacs.

The *amniotic sac* is lined with **ectoderm**. This lining comes to lie on the outer surface of the embryo.

The *yolk sac* is lined with **endoderm**. This comes to line the developing gut of the embryo.

The tissue between these two layers is **mesoderm**.

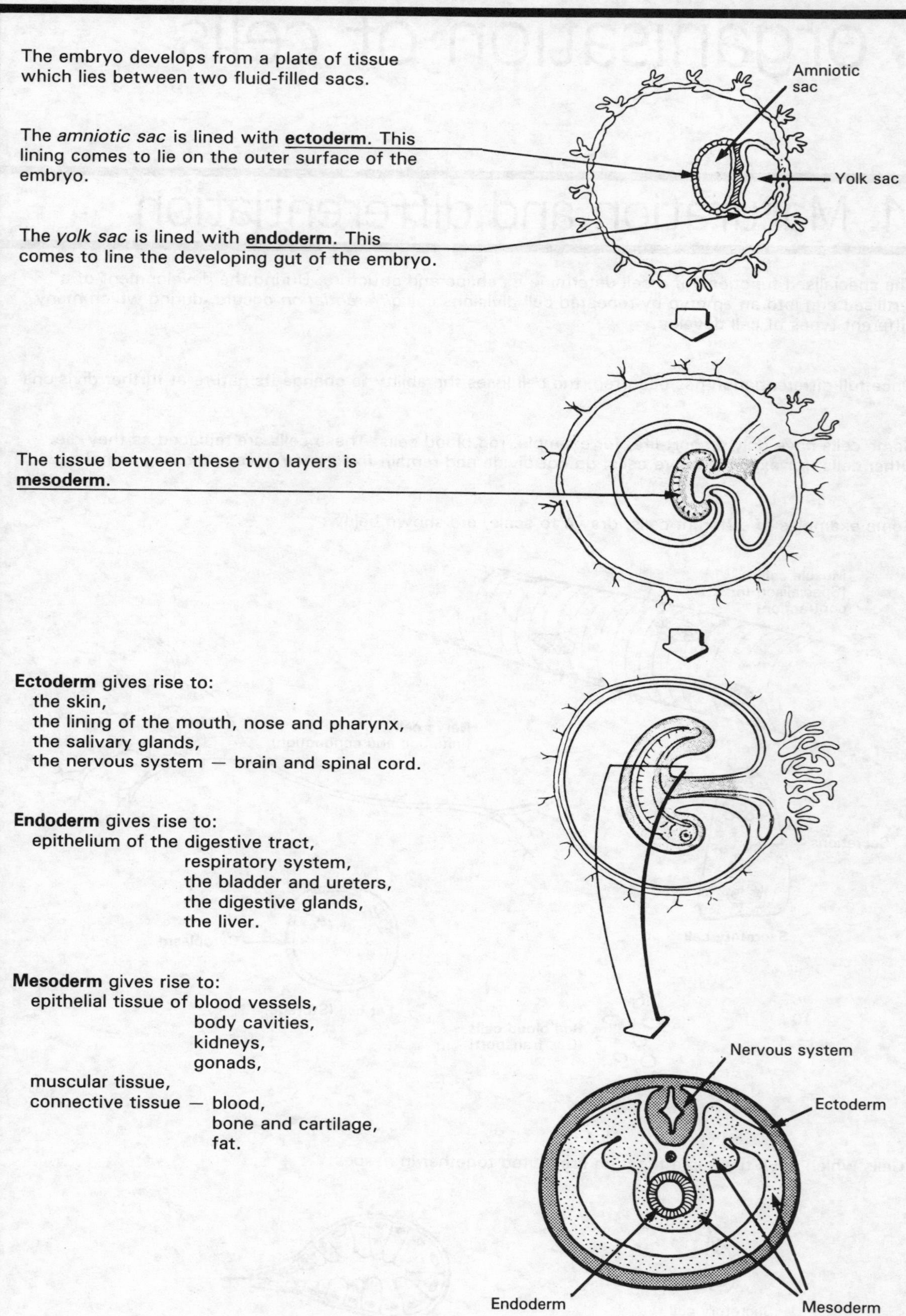

Ectoderm gives rise to:
- the skin,
- the lining of the mouth, nose and pharynx,
- the salivary glands,
- the nervous system — brain and spinal cord.

Endoderm gives rise to:
- epithelium of the digestive tract,
  - respiratory system,
  - the bladder and ureters,
  - the digestive glands,
  - the liver.

Mesoderm gives rise to:
- epithelial tissue of blood vessels,
  - body cavities,
  - kidneys,
  - gonads,
- muscular tissue,
- connective tissue — blood,
  - bone and cartilage,
  - fat.

TEST THREE

1. Indicate the functions for which the cells illustrated below are specialised, by placing ticks in the appropriate brackets.

		A	B	C
(a)	Contraction.	(✓)	( )	( )
(b)	Conduction.	( )	( )	(✓)
(c)	Storage.	( )	(✓)	( )

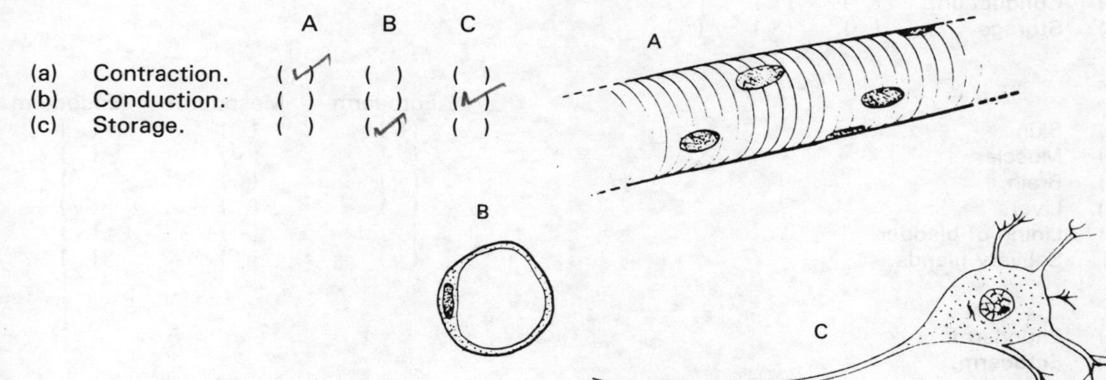

2. Indicate the embryonic layer from which the tissues listed below arise, by placing ticks in the appropriate brackets.

		Ectoderm	Mesoderm	Endoderm
(a)	Skin.	(✓)	( )	( )
(b)	Muscle.	( )	(✓)	( )
(c)	Brain.	(✓)	( )	( )
(d)	Liver.	( )	( )	(✓)
(e)	Lining of bladder.	( )	( )	(✓)
(f)	Salivary glands.	(✓)	( )	( )

3. (a) What name is given to the lining of the yolk sac? Endoderm

(b) What name is given to the lining of the amniotic sac? ectoderm

ANSWERS TO TEST THREE

1.

		A	B	C
(a)	Contraction.	(√)	( )	( )
(b)	Conduction.	( )	( )	(√)
(c)	Storage.	( )	(√)	( )

2.

		Ectoderm	Mesoderm	Endoderm
(a)	Skin.	(√)	( )	( )
(b)	Muscle.	( )	(√)	( )
(c)	Brain.	(√)	( )	( )
(d)	Liver.	( )	( )	(√)
(e)	Lining of bladder.	( )	( )	(√)
(f)	Salivary glands.	(√)	( )	( )

3. (a) Endoderm.
(b) Ectoderm.

4. Epithelial tissue and connective tissue

4.1. Epithelial tissue

Epithelial tissue is specialised to cover the surface of the body, the structures within the body, and to line the body's cavities.

It is protective in function and may be modified for absorption of selected substances, as in the gut and lung. It is capable of secretion, and secretory epithelium grows down into the underlying tissues to form a **gland**.

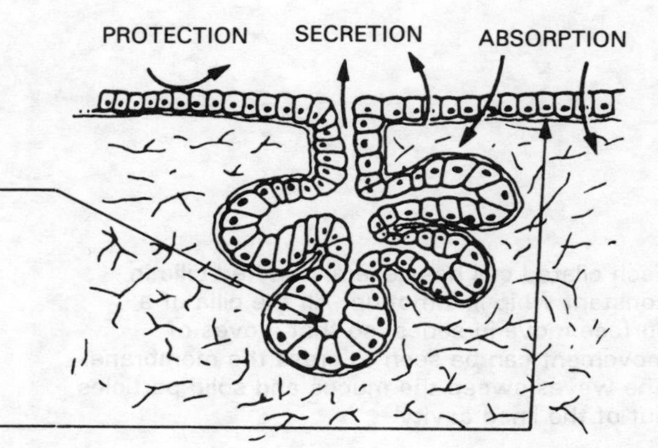

Epithelial tissue always lies on a **basement membrane.**

It may consist of a single layer of cells *(simple epithelium)* or many layers *(stratified epithelium).* Its cells are closely adherent to one another with no intracellular substance between them. It derives its nutrition from blood vessels which lie beneath the basement membrane.

Types of epithelial tissue:

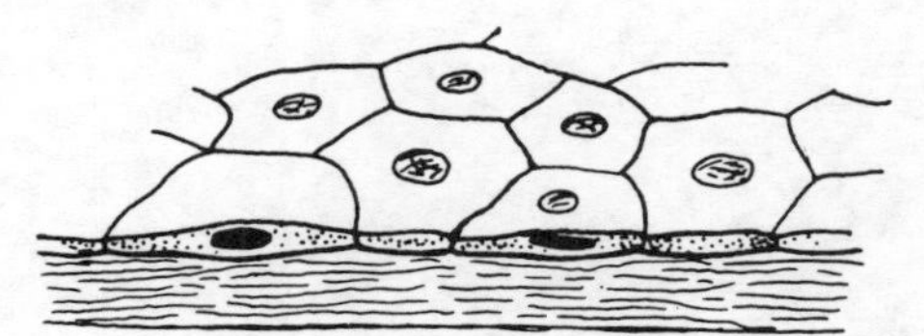

Squamous epithelium (blood vessels, body cavities)

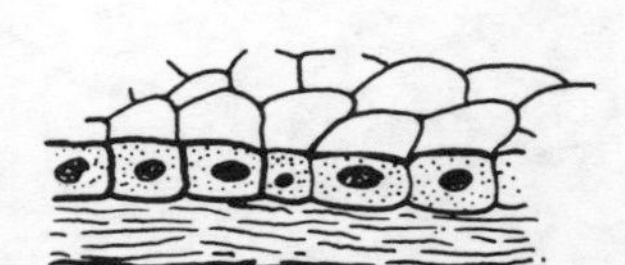

Cuboidal (surface of ovary)

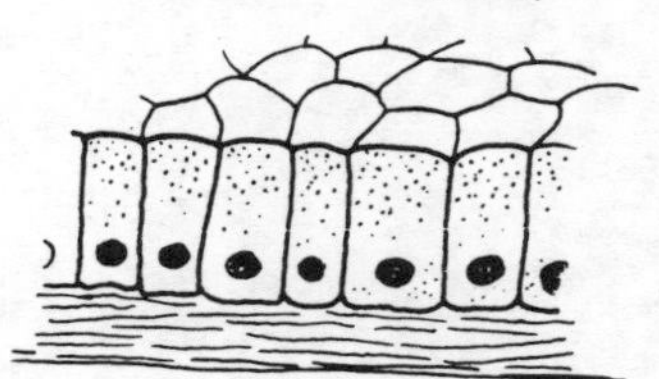

Columnar (cervix of uterus)

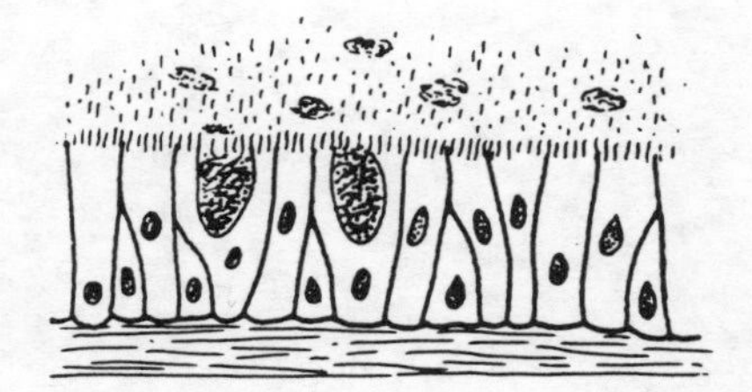

Ciliated columnar (trachea — windpipe)

Transitional (bladder)

Stratified (lining of mouth)

4.2. Ciliated epithelium

Some epithelial surfaces, for example those of the nasal and bronchial mucosa, have **cilia**. These are minute hairlike processes which serve to move mucus and solid particles. They are always interspersed with **goblet cells** which keep the surface moist with mucus.

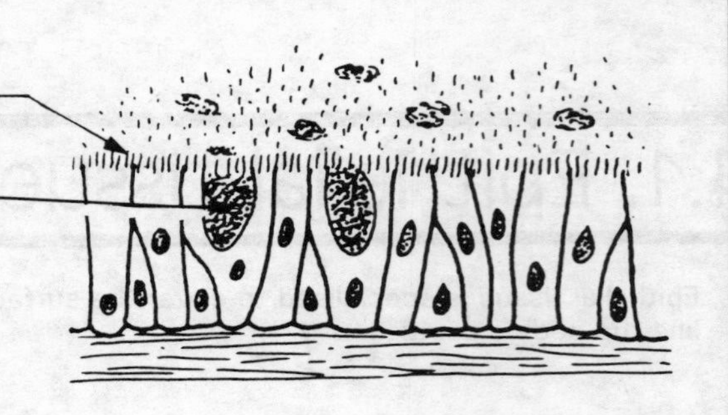

Each cilated cell has several hundred cilia in constant whip-like motion. All the cilia of a surface move in series, so that waves of movement can be seen to cross the membrane. The waves sweep the mucus and solid particles out of the lined cavity.

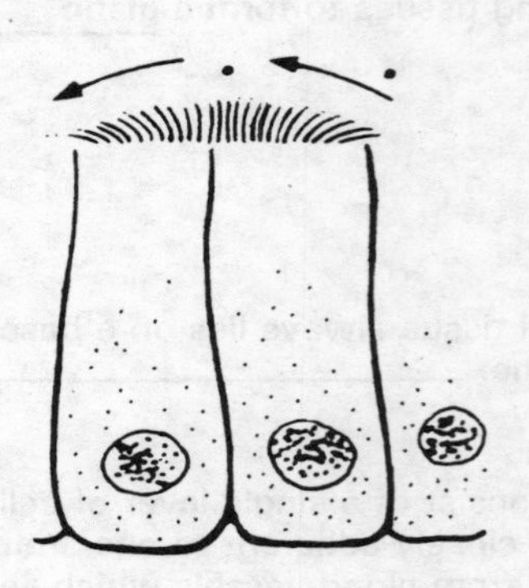

4.3. Secretory cells and glands

Some epithelial cells have highly developed powers of secretion.

They may be single secretory cells, scattered amongst non-secretory epithelial cells, for example the goblet cells in the intestinal epithelium, or they may be grouped together to form glands.

There are two types of gland.

Exocrine glands have ducts which carry the products of secretion to their local site of action, for example:

the **salivary glands** have ducts to the mouth,

the **sweat glands** have ducts to the skin surface.

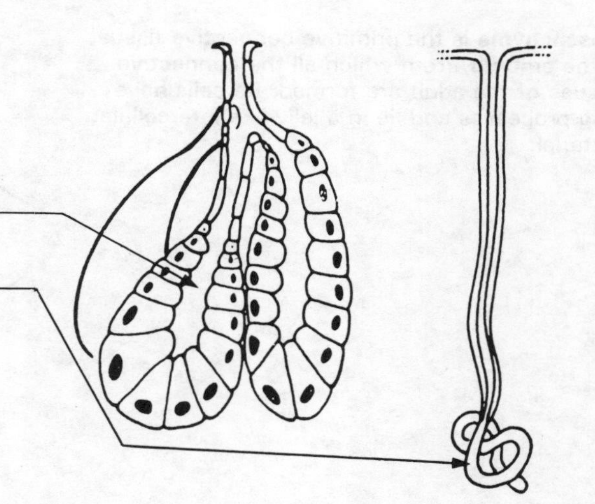

Endocrine glands do not have ducts. They have lost their connection with the surface from which they are developed and secrete directly into the bloodstream. The products of this secretion are transferred from the bloodstream to the entire body by the capillaries surrounding cells.

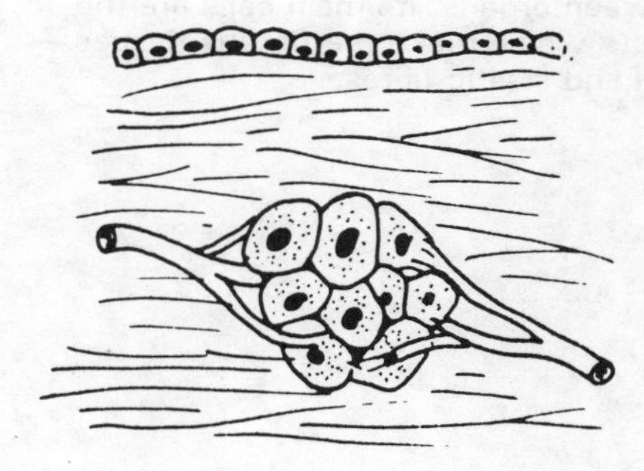

The secretions of endocrine glands are called *hormones.*

4.4. Connective tissue

Connective tissue develops from the mesoderm of the embryo. It lies between the other tissues giving them shape and support. It contains a great deal of non-living intercellular material which provides substance and strength.

Connective tissue also forms the transport system for the body (blood vessels etc.), and for the blood cells floating in the tissue.

Mesenchyme is the primitive connective tissue in the embryo, from which all the connective tissues of the adult are formed. Its cells have long processes and lie in a jelly-like intercellular material.

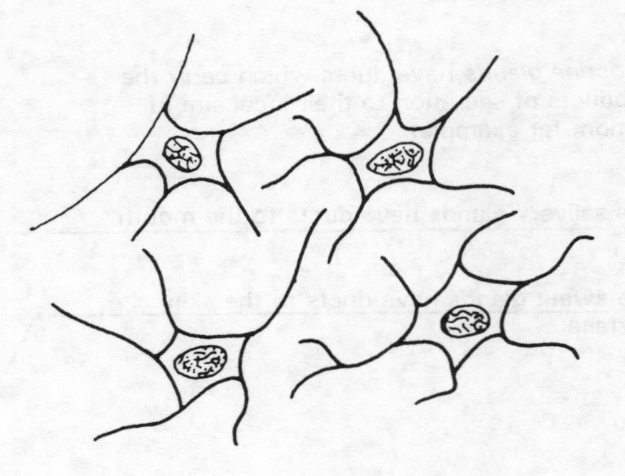

Loose areolar tissue is a packing tissue which lies between organs. Its main cells are the *fibroblasts* which make the fibrous tissues — collagen and elastic fibres.

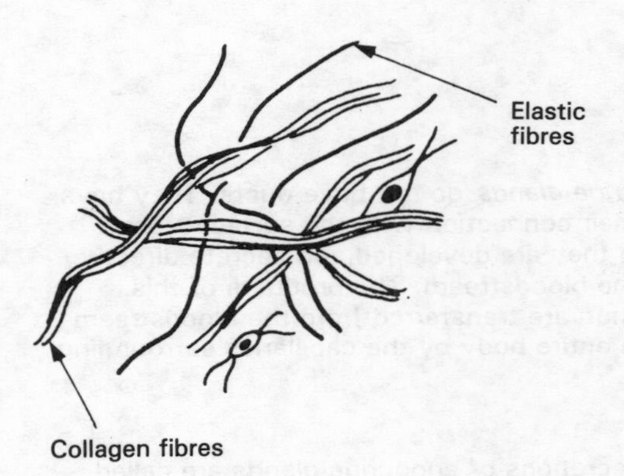

Dense fibrous tissue is formed by fibroblasts which create large amounts of collagen fibres. Collagen is a tough, white, inelastic protein material. Its fibres may be irregularly woven as in the heart valves and joint capsules, or they may be arranged as regular bundles all passing in one direction, as in tendons and ligaments.

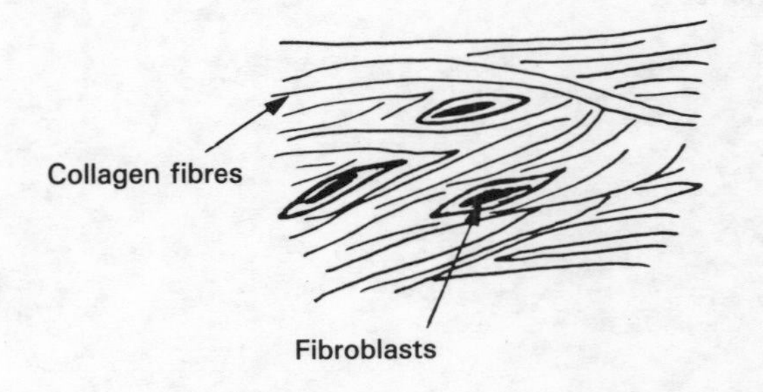

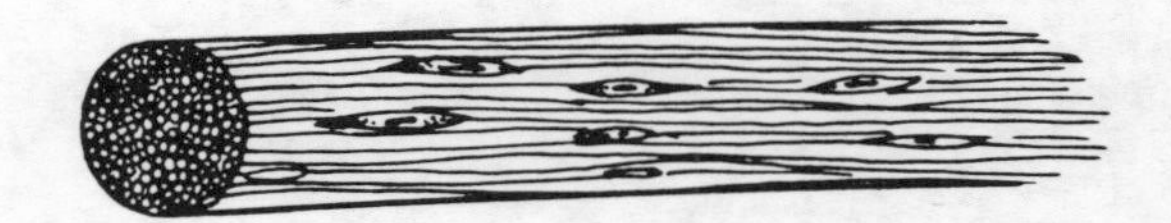

Elastic tissue, unlike tendons, can stretch. It is laid down as thin, corrugated plates. It is found mainly in the larger arteries.

Adipose tissue is a depot for fat reserves. It acts as a packing between organs and lies beneath the skin where it insulates against heat loss.

In the fat cells, the fat forms a single large storage droplet with a thin layer of protoplasm surrounding it.

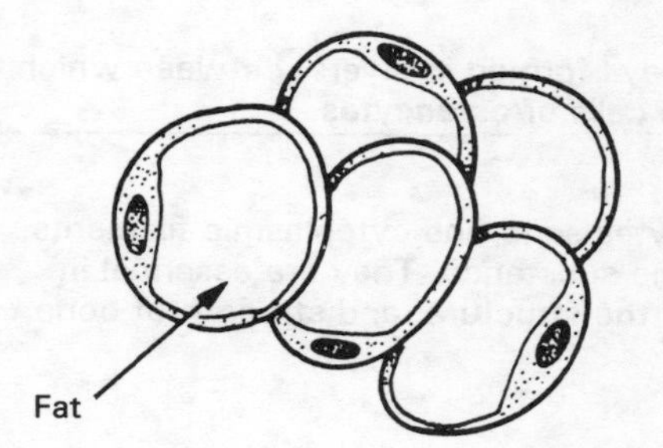

Cartilage is of three kinds.

Hyaline cartilage coats the surfaces of joints between bones, and forms the precursors of most bones in the embryo.

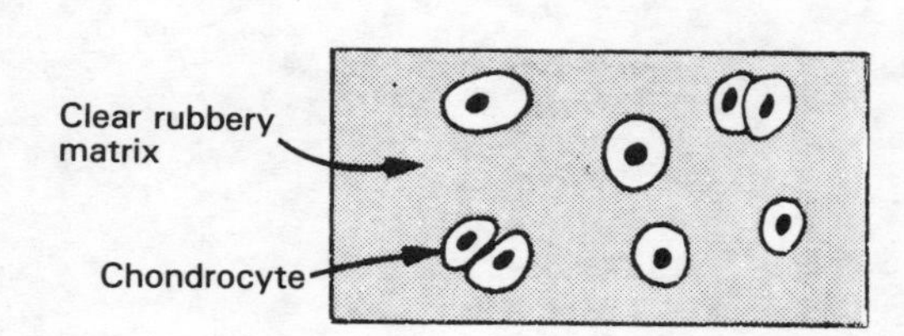

White fibrous cartilage is very hard, and found between the vertebrae.

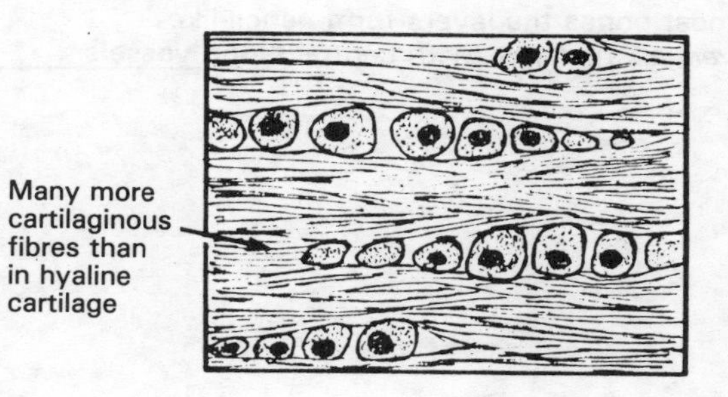

Elastic or yellow cartilage is flexible and elastic and is found in the larynx and external ear.

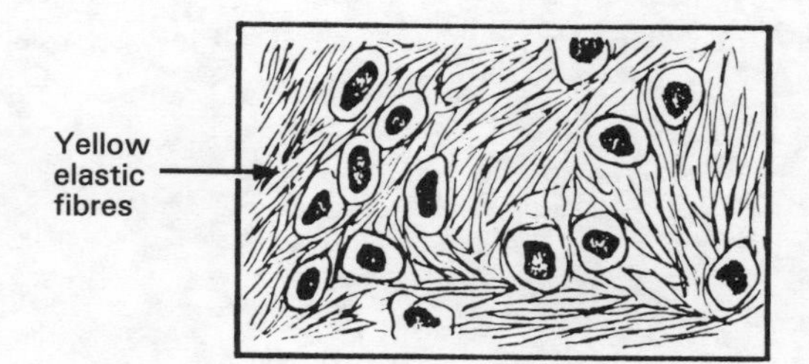

4.5. Bone

Bone is a living tissue with a rich blood supply. It can grow and repair itself when necessary.

Half of its mass is fibrous — a tough collagen framework. The other half is mineral, consisting of calcium salts deposited on the collagen fibres and making it rigid.

If a bone is heated in a furnace its fibrous part is destroyed and the remaining mineral part becomes hard, but brittle. If a bone is soaked in acid the mineral part dissolves away and the remaining fibrous part becomes tough and rubbery. The combinaton of these properties gives bone its unique rugged strength.

Bone is always formed in layers, between which lie the bone cells or **osteocytes**.

The osteocytes send fine cytoplasmic filaments into the bone substance. They are essential in maintaining the structure, and strength of bone.

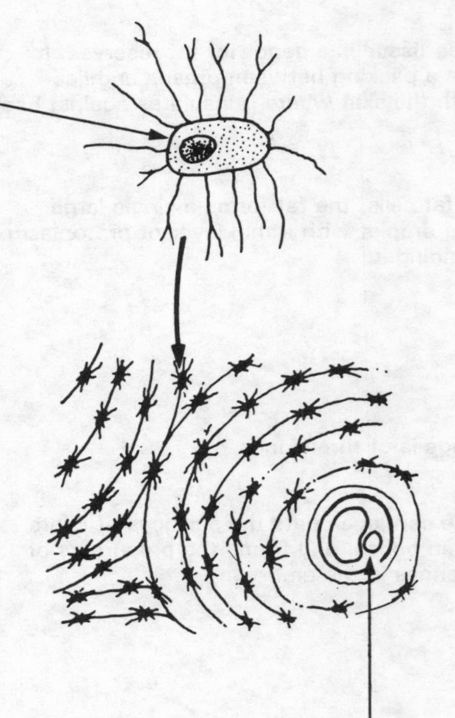

In most bones the layers form pencil-like *Haversian systems,* with central **blood vessels**.

4.6. Blood

The blood is made up of several different kinds of cells dispersed in a liquid plasma.

Red blood cells do not have nuclei. They are packed with the iron/protein complex called *haemoglobin*. They transport oxygen from the lungs to the tissues, and carbon dioxide from the tissues to the lungs.

Granulated white blood cells are of three main types:

— *neutrophils,* which engulf bacteria and digest them;

— *eosinophils,* which are involved in allergy reactions;

— *basophils,* the function of which is not completely understood.

Non-granulated white blood cells are of two types:

— *monocytes,* which destroy bacteria;

— *lymphocytes,* which are responsible for recognising material 'foreign' to the subject, and are responsible for the 'cell-mediated immune response'.

Platelets are non-nucleated fragments which play a part in clotting of the blood.

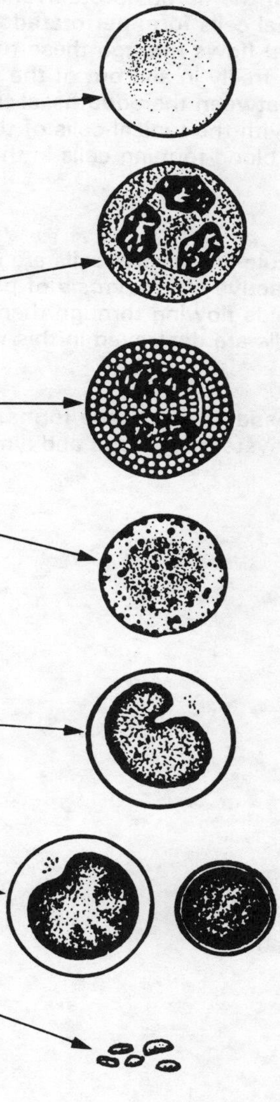

4.7. Reticulo-endothelial tissue

The blood-forming bone marrow, the spleen and the lymph nodes have similarities of structure. This structure consists of a network of fine threads of the fibrous protein *reticulin*, on which endothelial cells form perforated tubes. Blood (or lymph) flows through these tubes and cells can pass freely in and out of the tubes. The spaces between the endothelial tubes are packed with the typical cells of the tissue; for example blood-forming cells in the bone marrow.

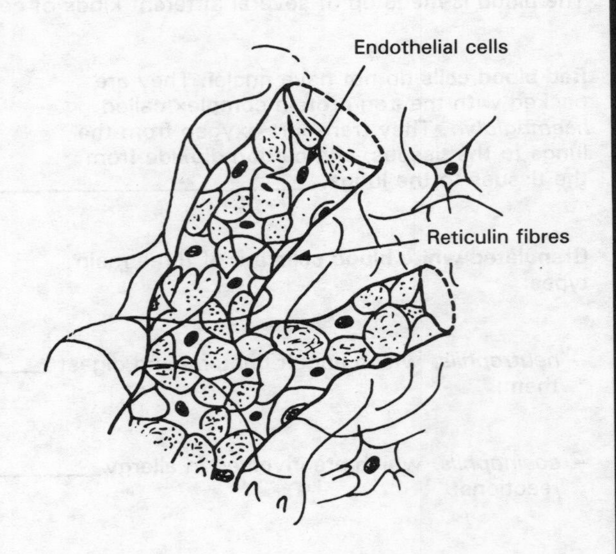

The reticulo-endothelial cells are characterised by their active *phagocytosis* of particles found in the fluids flowing through them. Degenerating blood cells are destroyed in this way.

Reticulo-endothelial tissue forms a complex filtering system for blood and lymph.

TEST FOUR

1. (a) What two features characterise ciliated epithelium?

   (b) Where is ciliated epithelium found?

2. Indicate which of the following are characteristics of endocrine glands, and which are characteristics of exocrine glands, by placing ticks in the appropriate brackets.

		Endocrine	Exocrine
(a)	Secretions have a mainly local action.	( )	( )
(b)	Secretions can affect the entire body.	( )	( )
(c)	Secretions are called hormones.	( )	( )
(d)	The glands secrete into ducts.	( )	( )
(e)	The glands secrete into the blood stream.	( )	( )

3. Which of the statements on the right apply to the tissues listed on the left?

(a)	Mesenchyme.	(i)	Forms a packing tissue between organs.
(b)	Loose areolar tissue.	(ii)	Largely composed of collagen.
(c)	Dense fibrous tissue.	(iii)	A primitive embryonic connective tissue.
(d)	Elastic tissue.	(iv)	Found in the larger arteries.

4. Which of the statements on the right apply to the blood cells listed on the left?

(a)	Lymphocytes.	(i)	Transport oxygen.
(b)	Red blood cells.	(ii)	Phagocytose bacteria.
(c)	Neutrophils.	(iii)	Recognise foreign material.

ANSWERS TO TEST FOUR

1. (a) Ciliated epithelium always has cilia, small hair-like organs, and goblet cells.
(b) The nose and the nasal section of the pharynx are lined with ciliated epithelial tissue.

2.

		Endocrine	Exocrine
(a)	Secretions have a mainly local action.	( )	(√)
(b)	Secretions can affect the entire body.	(√)	( )
(c)	Secretions are called hormones.	(√)	( )
(d)	The glands secrete into ducts.	( )	(√)
(e)	The glands secrete into the blood stream.	(√)	( )

3.

(a)	Mesenchyme	(iii)	is a primitive embryonic connective tissue.
(b)	Loose areolar tissue	(i)	forms a packing tissue between organs.
(c)	Dense fibrous tissue	(ii)	is largely composed of collagen.
(d)	Elastic tissue	(iv)	is found in the larger arteries.

4.

(a)	Lymphocytes	(iii)	recognise foreign material.
(b)	Red blood cells	(i)	transport oxygen.
(c)	Neutrophils	(ii)	phagocytose bacteria.

5. Muscular tissue, nervous tissue and skin

5.1. Muscular tissue

There are three types of muscular tissue.

Smooth muscle *(involuntary muscle)* consists of elongated spindle-shaped cells about 0.2 mm long and lying parallel to each other. Each cell has a single nucleus.

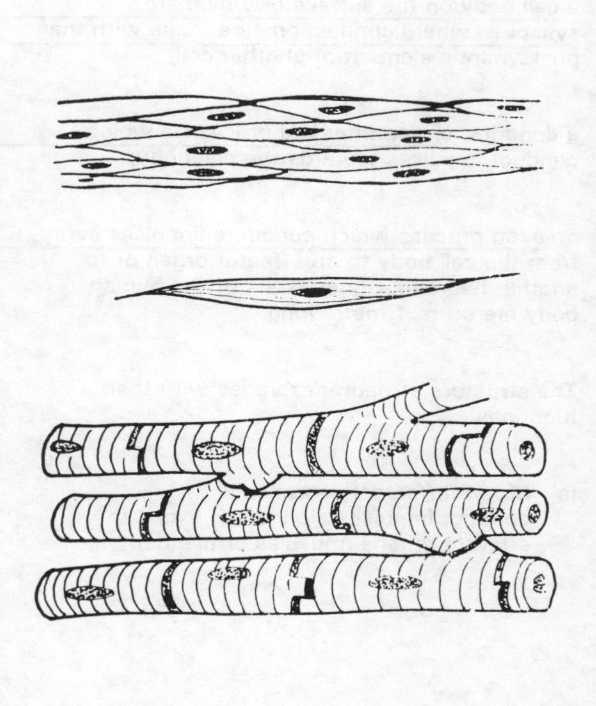

Smooth muscle is under the control of the autonomic nervous system but also possesses its own inherent rhythmic contraction. It is found in the uterus, blood vessels, gut, ureters and bladder.

Cardiac muscle is found only in the heart. The individual cells have a nucleus each and the branches of the cells are strongly adherent to each other.

Cardiac muscle has its own rhythmic contractions, normally controlled by a bundle of neuromuscular cells, the sinuatrial node.

Cardiac muscle has a rich blood supply. It can be seen to have a cross-striped pattern, similar to that in skeletal muscle.

Striated *(skeletal)* **muscle** forms the bulky muscles of the limbs and trunk. It accounts for about half the weight of the adult body. It is formed from very long multinucleated cells which show a striking striated pattern when examined under the microscope.

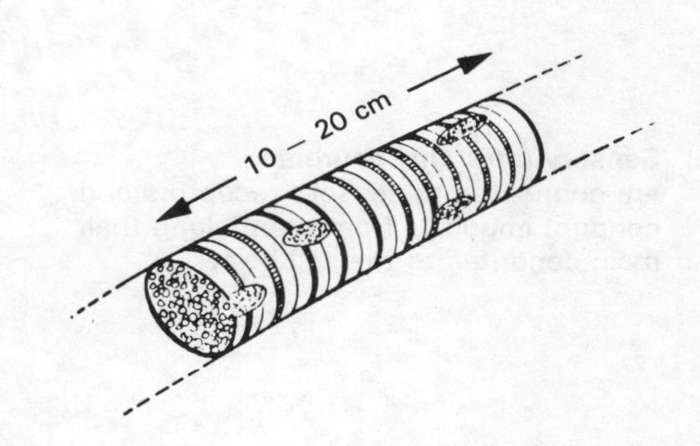

Skeletal muscle is capable of much more rapid contraction than the other muscles. It is under the complete control of the nerves which supply it and wastes away if its nerve is cut.

The muscle has a rich blood supply which provides oxygen and nutrients for its activity. It also contains glycogen, a sugar, as an emergency supply of food, and myoglobin which stores oxygen ready for immediate use. Myoglobin is similar to the blood protein haemoglobin and gives muscle its characteristic red colour.

5.2. Nervous tissue

Nervous tissue is formed from *neurones*, which are cells with sometimes extremely long processes specialised for the conduction of nerve impulses, and *glial cells,* which support and nourish the neurones.

Neurones consist of:

a **cell body** on the surface of which are **synapses** where connections are made with the postsynaptic element of another cell;

a **dendrite**, or **dendrites** are processes which conduct impulses towards the cell body;

an **axon** process which conducts impulses away from the cell body to an effector organ or to another neurone. Some axons of the human body are up to 1 metre long.

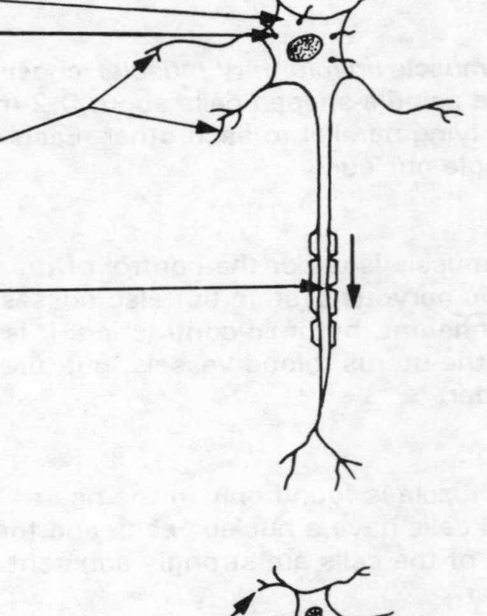

The structure of neurones varies with their functions.

(a) **Motor** ***(efferent)*** **neurones**
conduct impulses along their axons to effector organs (muscles or glands).

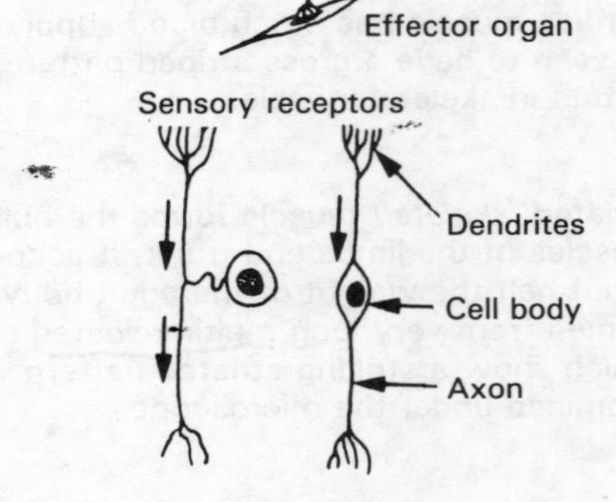

(b) **Sensory** ***(afferent)*** **neurones**
are connected to sensory receptors and conduct impulses from them along their main dendrites to the cell body.

(c) **Association neurones**
relay impulses between other neurones (e.g. the cells of the cerebral cortex of the brain and in the spinal cord.)

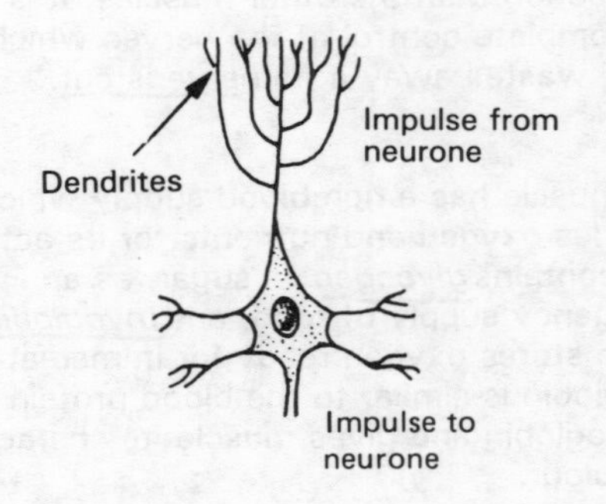

5.3. Skin

The skin has two main layers, the *dermis* and the *epidermis*.

The epidermis is stratified squamous epithelium and is made up of:

an outer **stratum corneum** composed of dead cells which have been converted to keratin;

several intermediate layers where conversion to keratin is occurring;

a **germinal layer** where new cells are constantly being produced. Pigment-producing cells *(melanocytes)* lie among the germinal cells.

The **dermis** lies beneath the epidermis. It is tough and thick. It is a felt-work of collagen fibres with a very rich blood and nerve supply.

It contains **hair follicles**
sweat glands
and subcutaneous glands.

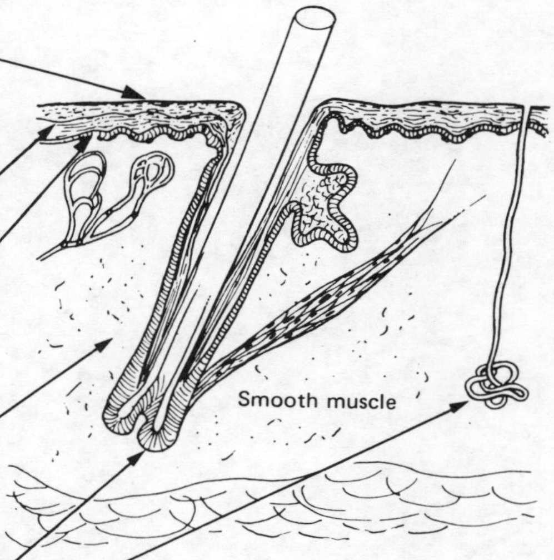

Smooth muscle

Beneath the dermis lies a layer of adipose tissue called the *superficial fascia.*

The functions of the skin are:

(1) protection, against injury, and bacteria;
(2) regulation of body temperatures through variations in the flow of blood through the dermis and subcutaneous tissue, and evaporation of sweat secreted onto the surface by the sweat glands;
(3) excretion (slight) of water and some salt through the sweat glands;
(4) acting as a sense organ for touch, pain, heat and cold.

TEST FIVE

1. Indicate which of the following are characteristics of the different types of muscle, by placing ticks in the appropriate brackets.

		Smooth	Cardiac	Striated
(a)	Cells form a branching system.	( )	(✓)	( )
(b)	Found in the uterus.	(✓)	( )	(✓)
(c)	Formed of multinucleated cells.	( )	(✓)	( )
(d)	Cells strongly adherent to each other.	( )	( )	( )
(e)	Wastes away when its nerve supply is cut.	(✓)	( )	( )

2. Indicate which of the items in the list below apply to the parts of the cell illustrated in the diagram, by placing the appropriate letters in the brackets.

1. Dendrite. (A)
2. Axon. (D)
3. Nucleus. (B)
4. Postsynaptic membrane. (C)

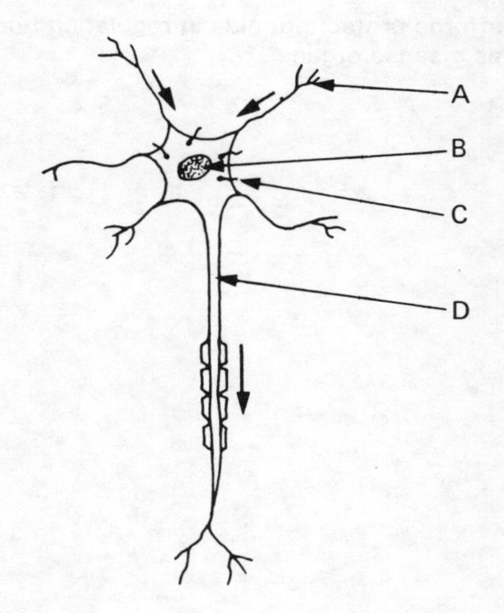

3. What are the functions served by the skin?

Protection, Temperature regulation
sense organ for touch, heat, pain cold

ANSWERS TO TEST FIVE

1.

		Smooth	Cardiac	Striated
(a)	Cells form a branching system.	( )	(√)	( )
(b)	Found in the uterus.	(√)	( )	( )
(c)	Formed of multinucleated cells.	( )	( )	(√)
(d)	Cells strongly adherent to each other.	( )	(√)	( )
(e)	Wastes away when its nerve supply is cut.	( )	( )	(√)

2.

1. Dendrite. (A)
2. Axon. (D)
3. Nucleus. (B)
4. Postsynaptic membrane. (C)

3. The skin affords protection, aids in regulating body temperature, aids in excreting salt and moisture, and acts as a sense organ.

6. The human body

6.1. Cells, tissues, organs and systems

A *tissue* is a group of cells performing similar physiological functions.

An *organ* is a structural and functional unit formed from several different tissues and performing a highly specialised function.

For example, the **stomach** is an organ specialising in the temporary storage and initial digestion of food. It is made up of all four types of tissue — epithelial (inner lining and outer coat), connective and muscular and nervous tissue (in walls of stomach).

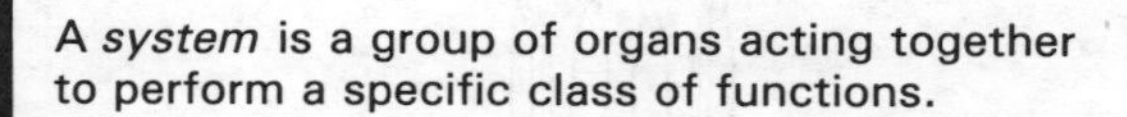

A *system* is a group of organs acting together to perform a specific class of functions.

For example, the digestive system includes the following organs:

— the **oesophagus**

— the **stomach**

— the **small intestine**

— the **large intestine**

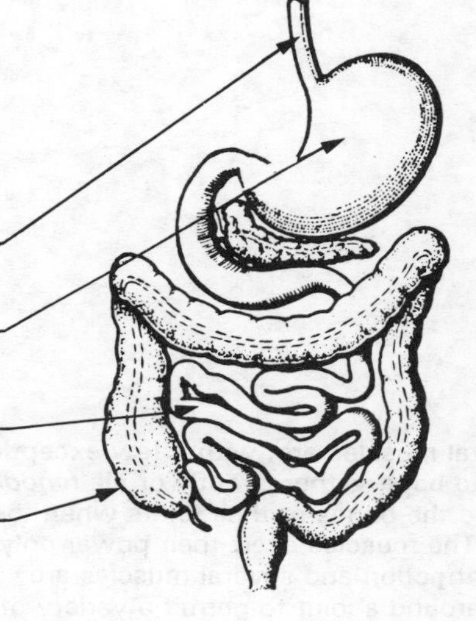

and is responsible for digestion and absorption of all the food taken into the body.

The following are the body's systems:

(1) the *nervous system,* including the senses;

(2) the *locomotor system* (the skeleton and skeletal muscles);

(3) the *endocrine system* (the ductless glands);

(4) the *reproductive system*;

(5) the *cardiovascular system* (the heart and blood vessels);

(6) the *respiratory system* (the lungs and air passages);

(7) the *digestive system*;

(8) the *urinary system* (the kidneys and bladder).

6.2. Movement

The skeleton provides a framework to support the body and maintain its shape.

In places it surrounds and protects the soft internal organs.

The main function of the skeleton is to provide a system of levers moved by skeletal muscle, allowing the body to move.

The individual bones are attached to one another by complex **joints** most of which allow a variable degree of movement between the bones and yet are strong enought to prevent the bones being pulled apart.

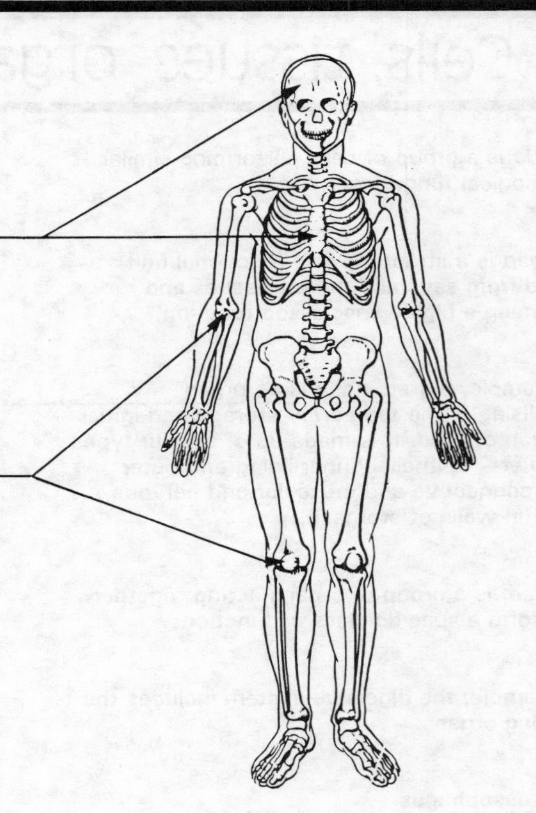

The skeletal muscles are, with a few exceptions attached to bone, either directly or via *tendons*. They move the bones at their joints when they contract. The muscles exert their power only during contraction and several muscles are attached around a joint to permit a variety of opposing movements.

Skeletal muscle makes up about half the weight of the average adult.

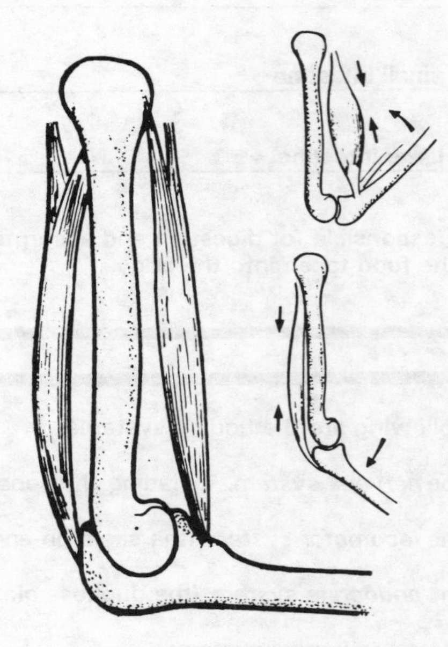

The elbow joint

6.3. Control and co-ordination

If movements are to be meaningful, information must be collected about the environment and the relation of the body to it. This is the function of the senses.

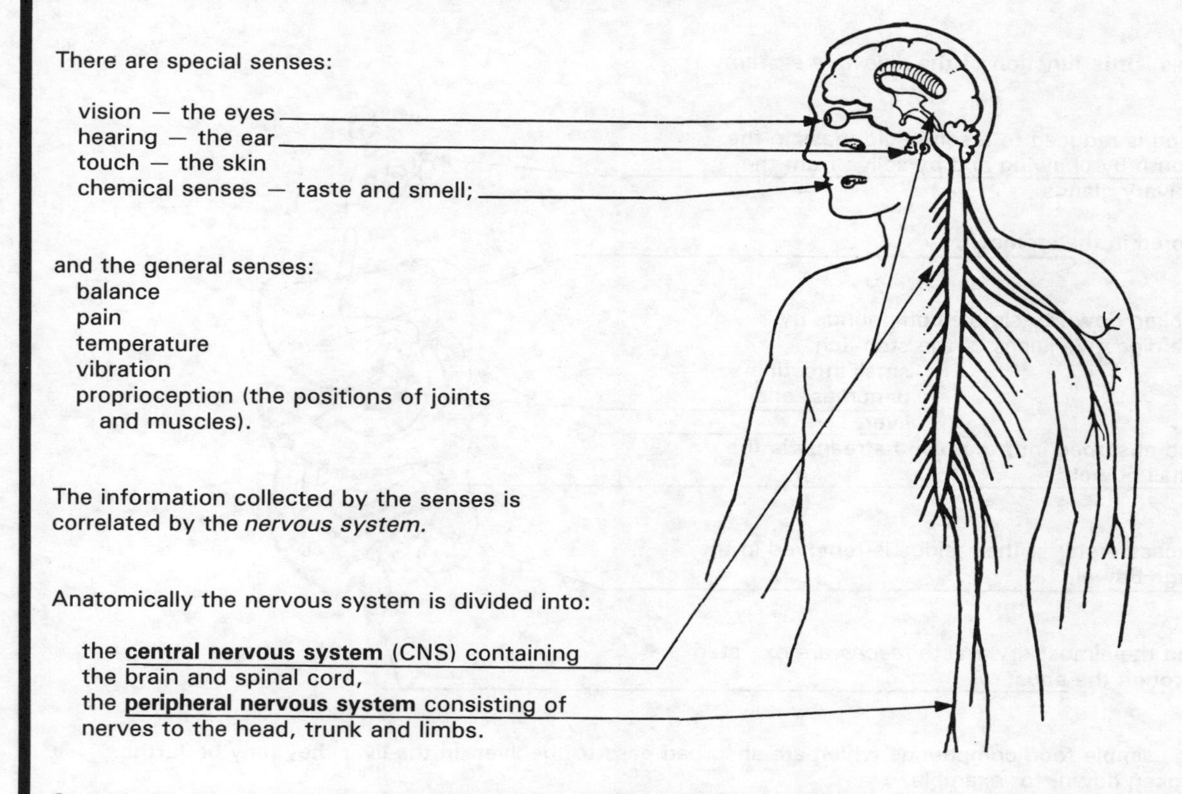

There are special senses:

vision — the eyes
hearing — the ear
touch — the skin
chemical senses — taste and smell;

and the general senses:
balance
pain
temperature
vibration
proprioception (the positions of joints and muscles).

The information collected by the senses is correlated by the *nervous system.*

Anatomically the nervous system is divided into:

the **central nervous system** (CNS) containing the brain and spinal cord,
the **peripheral nervous system** consisting of nerves to the head, trunk and limbs.

Control of movement is a function of the brain, which analyses incoming information and can cause accurate movements with controlled contraction of the skeletal muscles, so that the body can rapidly respond to changes in the environment or to the needs of the individual.

The CNS also plays a part in the control of the endocrine glands and some exocrine glands.

6.4. Nutrition

In order to provide the raw materials and energy for cellular activities food must be eaten, broken down into the few simple forms which the body is able to utilise, and absorbed into the body.

This is the function of the *digestive system*.

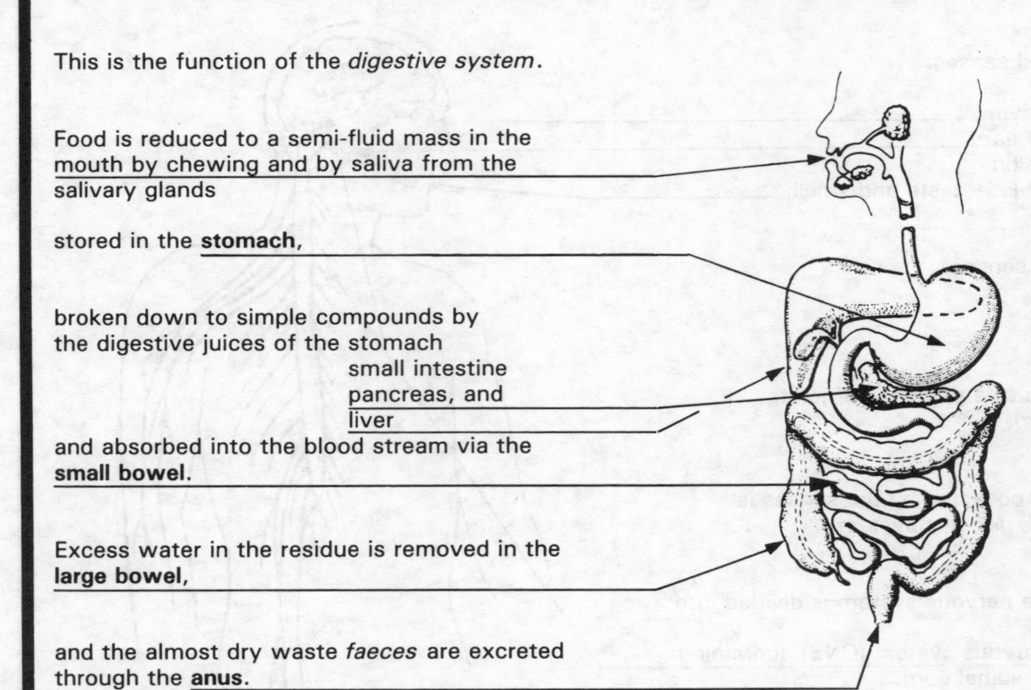

The simple food compounds which are absorbed pass to the liver. In the liver they may be further broken down, for example:

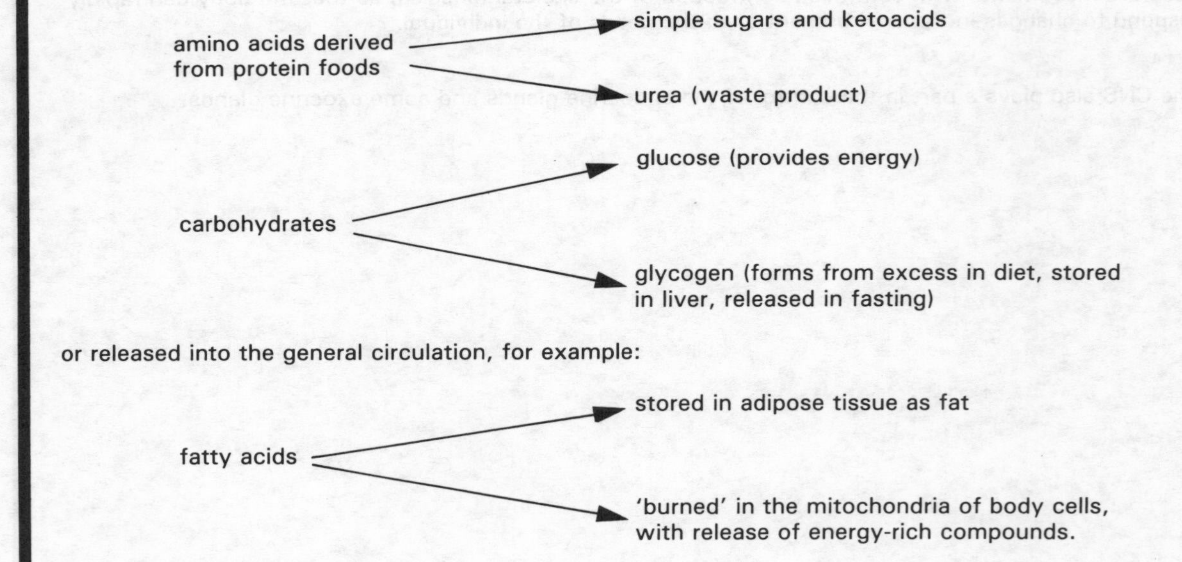

6.5. Transport and distribution

The body cells need a constant supply of vital materials such as glucose, amino acids and oxygen. The waste products of metabolism (nitrogen compounds, acids, carbon dioxide, heat) must also be constantly removed from the cells.

These functions are carried out by the constant circulation of a complex fluid, *blood*, through the tissue of the body.

This circulation is powered by the **heart**, a muscular pump, which forces blood at *high pressure* into a closed, branching system of tubes, the **arteries**.

The blood is distributed to all parts of the body.

In the tissue the blood flows through thin-walled vessels called **capillaries**. Here exchange of nutrients, gases and waste products occurs.

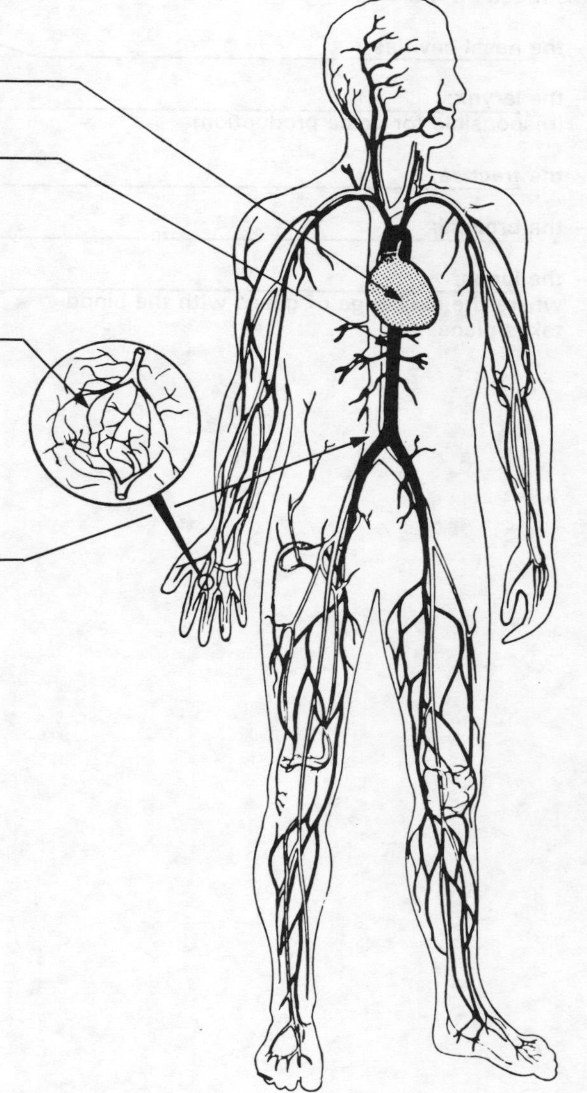

The blood then passes into a *low pressure* collecting system of vessels, the **veins**. The deep veins run parallel to the arteries and return the blood to the heart.

Before returning to the general circulation the blood is first pumped through the lungs to pick up fresh oxygen and excrete its carbon dioxide.

The adult human heart pumps about 5 litres of blood per minute at rest. Its output rises to over 20 litres per minute during exercise.

6.6. Respiration

The respiratory system is responsible for introducing oxygen into the blood, which carries it to the tissues, and for removing carbon dioxide from the blood.

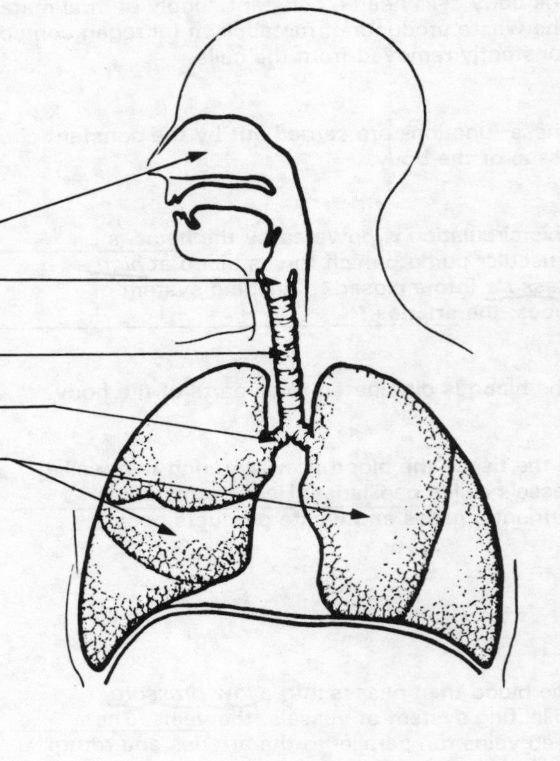

It is made up of:

— the **nasal cavities**;

— the **larynx**;
(responsible for voice production)

— the **trachea**;

— the **bronchi**;

— the **lungs**;
where the exchange of gases with the blood takes place.

6.7. Excretion

Nitrogenous wastes (mainly as the simple compound *urea*) and acids are removed from the blood by the *kidneys*.

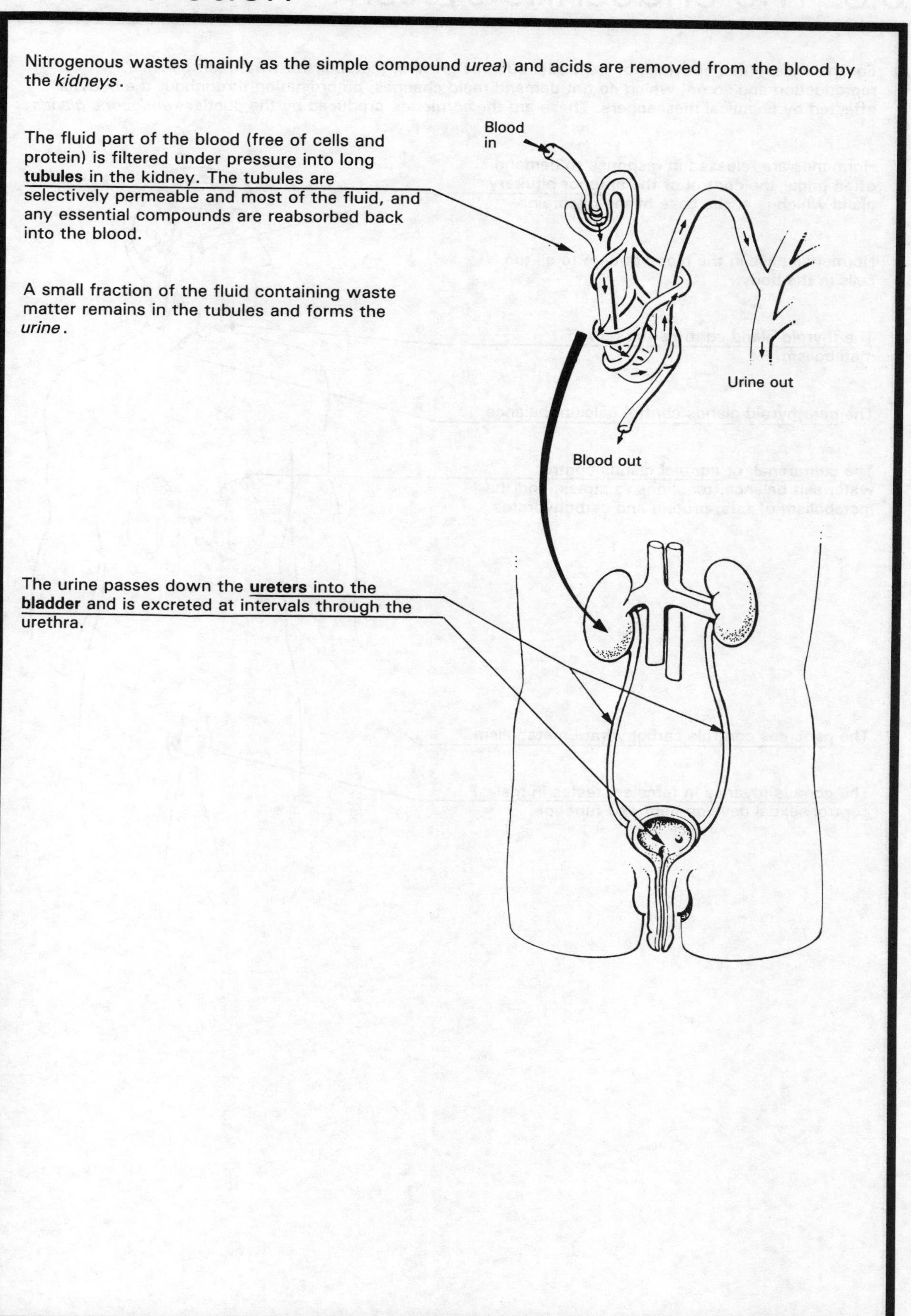

The fluid part of the blood (free of cells and protein) is filtered under pressure into long **tubules** in the kidney. The tubules are selectively permeable and most of the fluid, and any essential compounds are reabsorbed back into the blood.

A small fraction of the fluid containing waste matter remains in the tubules and forms the *urine*.

The urine passes down the **ureters** into the **bladder** and is excreted at intervals through the urethra.

6.8. The endocrine system

For the widespread activities of the body, such as growth, metabolism, water/salt balance, reproduction and so on, which do not demand rapid changes, co-ordination throughout the body is effected by chemical messengers. These are the *hormones* produced by the ductless *endocrine glands*.

Hormones are released in response to demand, often under the control of the **anterior pituitary gland** which is at the base of the midbrain.

Hormones pass in the blood stream to all the cells in the body.

The **thyroid gland** controls the rate of metabolism.

The **parathyroid glands** control calcium balance.

The **suprarenal**, or **adrenal** glands control water/salt balance, reactions to stress, and the metabolism of fats, protein and carbohydrates.

The **pancreas** controls carbohydrate metabolism.

The **gonads** (ovaries in females, testes in males) control sexual development and function.

6.9. Reproduction

To continue the species, a mechanism is needed for the production of germ cells in both sexes, their transfer to the female to permit fertilisation and the subsequent development of the embryo within a protected environment in the female.

Each month between the cyclical bleeds from the womb *(menstrual periods)* an egg or **ovum** is released from the ovary *(ovulation)*. The egg is picked up by the **uterine tube**.

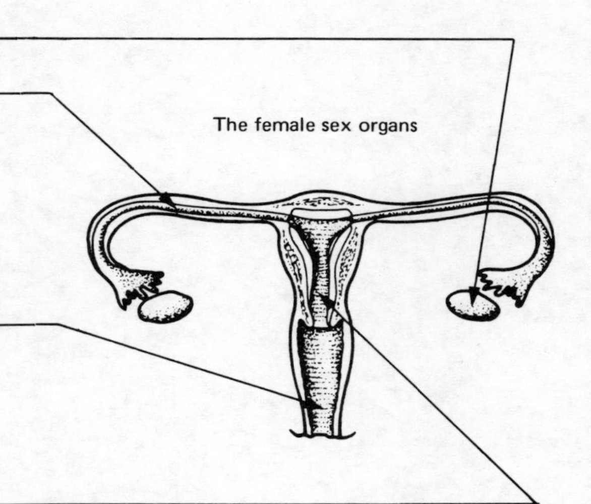

In the tube it may be fertilised by a single *spermatozoon*, the male germ cell.

Spermatozoa are developed in the testes of the male. Several hundred million spermatozoa are deposited in the upper part of the **vagina** during sexual intercourse. They swim through the neck of the womb, through the cavity of the womb, and into the uterine tubes.

If fertilisation does occur, the fertilised egg passes into the cavity of the **uterus** where it sinks into the thick lining prepared for it.

An embryo develops in the uterus which enlarges to contain it. The developed baby is expelled by powerful contractions of the smooth muscle in the wall of the uterus about 38 weeks after fertilisation.

If fertilisation does not occur the lining of the uterus is shed 14 days after ovulation. This shedding is the menstrual period.

TEST SIX

1. What is the connection between tissue and cells and tissue and organs?

A tissue is a collection of cells with the same function

A organ may have different tissues collected tog to perform a specific funct

2. What is a *system* of the body?

several organs performing a function eg digestion

3. Does the skeleton protect the central nervous system?

Yes skull surrounds brain Backbone surrounds spinal cord

4. Which of the statements on the right apply to the blood vessels listed on the left?

(a)	Arteries. ii	(i)	Responsible for exchange of oxygen and nutrients.
(b)	Capillaries. i	(ii)	Responsible for high pressure distribution of blood.
(c)	Veins. iii	(iii)	Responsible for low pressure return flow of blood.

5. Indicate which of the items in the list below apply to the glands labelled on the diagram alongside, by placing the appropriate letters in the brackets.

1. Gonads. (D)
2. Thyroid. (B)
3. Suprarenals. (C)
4. Pituitary. (A)

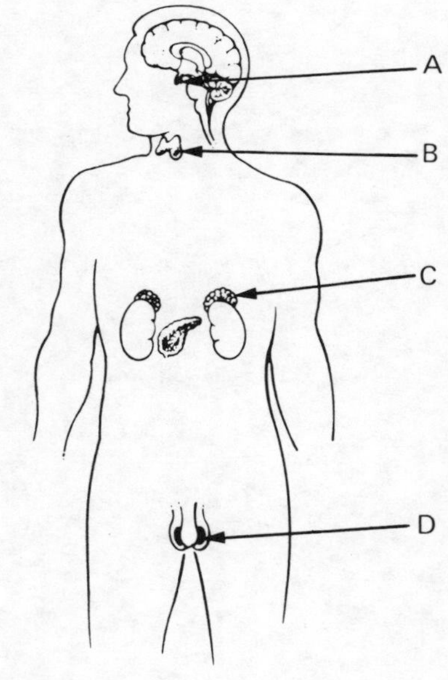

ANSWERS TO TEST SIX

1. Tissue is made up of cells of the same type, and an organ is made up of different types of tissue that together enable performance of a specific function.

2. A system is a group of organs which act together to perform a specific class of functions.

3. Yes, the skull covers and protects the brain, the vertebral column surrounds and protects the spinal cord.

4.

(a)	Arteries are	(ii)	responsible for high pressure distribution of blood.
(b)	Capillaries are	(i)	responsible for exchange of oxygen and nutrients.
(c)	Veins are	(iii)	responsible for low pressure return flow of blood.

5.

1.	Gonads.	(D)
2.	Thyroid.	(B)
3.	Suprarenals.	(C)
4.	Pituitary.	(A)

POST TEST

1. Which of the statements on the right apply to the features of a cell listed on the left?

(a)	Mitochondria. IV	(i)	Associated with secretion.
(b)	Golgi apparatus. I	(ii)	Involved in protein synthesis.
(c)	Ribosomes. II	(iii)	Contains the cell's DNA.
(d)	Centrosomes. V	(iv)	The site of energy production.
(e)	The nucleus. III	(v)	Involved in cell division.

2. (a) Name two bases found in DNA. Guanine cytosine
(b) How many bases are there in the base sequence for one amino acid in a strand of DNA? 3

3. Which of the processes listed below involve activity of the cell membrane?

(a) Osmosis. ✓
(b) Phagocytosis. (b)
(c) Secretion.
(d) Diffusion.

4. Indicate in which order the drawings below should be, to show the correct sequence of events in mitosis.

1. (B)
2. (E)
3. (D)
4. (A)
5. (C)

A B C

D E

5. How does the endocrine system affect the body's functions?

Endocrine glands release hormones into bld affecting all systems

POST TEST

6. Indicate which of the drawings below are described by the labels listed on the left, by placing the appropriate letters in the brackets.

1. Ciliated columnar epithelium. (D)
2. Cuboidal epithelium. E (B)
3. Squamous epithelium. (A)
4. Transitional epithelium. (C)
5. Columnar epithelium. (B)

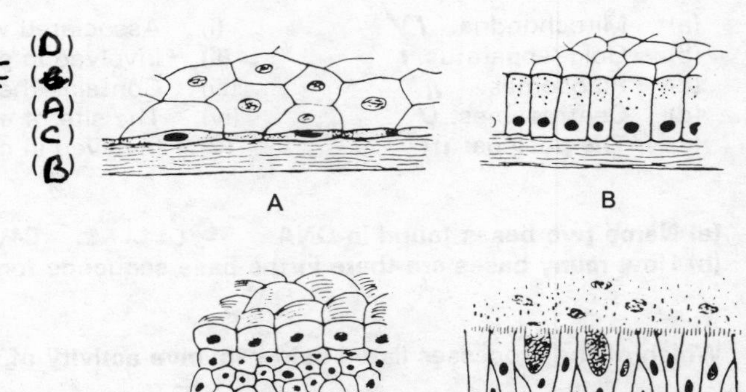

C D

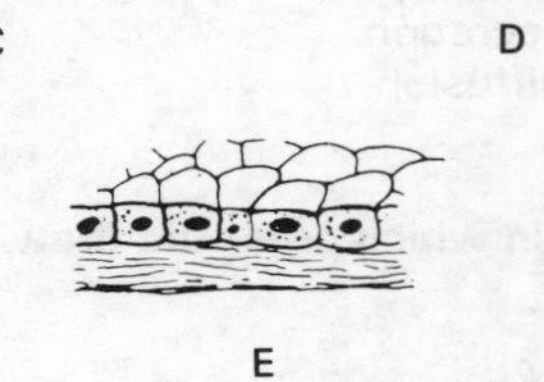

7. Which of the blood cells listed on the left are associated with the functions listed on the right?

(a)	Red blood cells. iv	(i)	Phagocytosis of bacteria.
(b)	Platelets. iii	(ii)	Formation of antibodies.
(c)	Neutrophils. i	(iii)	Coagulation of blood.
(d)	Eosinophils.	(iv)	Transport of oxygen.
(e)	Monocytes. ii		
(f)	Lymphocytes.		

8. Which of the following structures contain large amounts of collagen?

(a) Tendons. (b) Bones. (c) Large arteries. (d) Cartilage. (e) Heart valves.

9. Complete the following:

(a) Skeletal muscle is attached to bone either directly or via tendons
(b) ________refers to the position of joints and muscles. proprioception
(c) The heart is a organ muscular pump
(d) The digestive juices responsible for breaking down the contents of the stomach are found in the stomach, the pancreas, the small intestine, and the liver.

10 (a) How much blood per minute does the heart pump round the body at rest? 5 L

(b) Through which organ does the heart pump all the blood with each complete circulation?

Lungs

ANSWERS TO POST TEST

1. (a) Mitochondria (iv) are the site of energy production.
   (b) Golgi apparatus (i) is associated with secretion.
   (c) Ribosomes (ii) are involved in protein synthesis.
   (d) Centrosomes (v) are involved in cell division.
   (e) The nucleus (iii) contains the cell's DNA.

2. (a) Two of the following:
   Adenine.
   Thymine.
   Guanine.
   Cytosine.

   (b) Three.

3. (b) Phagocytosis.
   (c) Secretion.

4. 1. (B)
   2. (E)
   3. (D)
   4. (A)
   5. (C)

5. The endocrine system is made up of hormone-secreting endocrine glands, and hormones are the messengers enabling the body to co-ordinate such activities as growth, metabolism, reproduction, digestion and blood pressure.

ANSWERS TO POST TEST

6. 1. Ciliated columnar epithelium. (D)
2. Cuboidal epithelium. (E)
3. Squamous epithelium. (A)
4. Transitional epithelium. (C)
5. Columnar epithelium. (B)

7. (a) Red blood cells (iv) are associated with oxygen transport.
(b) Platelets (iii) are associated with coagulation of blood.
(c) Neutrophils (i) are associated with phagocytosis of bacteria.
(d) Eosinophils are associated with none of the functions listed.
(e) Monocytes (i) are associated with phagocytosis of bacteria.
(f) Lymphocytes (ii) are associated with formation of antibodies.

8. (a) Tendons. (b) Bones. (e) Heart valves.

9. (a) Skeletal muscle is attached to bone either *directly* or *via tendons.*
(b) *Proprioception* refers to the position of joints and muscles.
(c) The heart is a *muscular pump*.
(d) The digestive juices responsible for breaking down the contents of the stomach are found in the *stomach*, the *small intestine*, the *pancreas*, and the *liver*.

10 (a) On average 5 litres,
(b) The lungs.

Contents

The Reproductive System

1. The male reproductive system

1.1 Anatomy and physiology

The primary sex organs in the male are the two testes. They lie in the scrotum outside the body cavities where they can be kept cool, as they need to be if they are to function fully.

Each testis consists of about 200 tightly coiled **seminiferous tubules** which lie within fibrous compartments. Each tubule is about 60 cm long. The male sex cells, the spermatozoa, are produced within the tubules, and emptied into their main duct, the **vas deferens**, via the long coiled **duct of the epididymis.**

The vas deferens carries the spermatozoa from the testis through the inguinal canal over the pubic bone, and around the side wall of the pelvis to the back of the bladder.

Here the spermatozoa can enter the urethra through the narrow **ejaculatory ducts.**

The **prostate**,

the **seminal vesicles**, and

the **bulbo-urethral glands**

add their secretions which form 90% of the volume of the semen.

The penis consists of two **corpora cavernosa**, which diverge to become attached to the pelvis, and a **corpus spongiosum** which carries the urethra. The corpus spongiosum expands to form the **glans** at the tip of the penis.

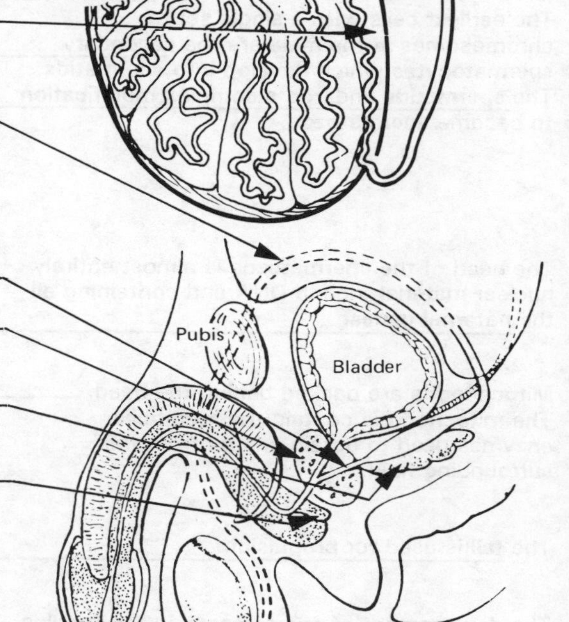

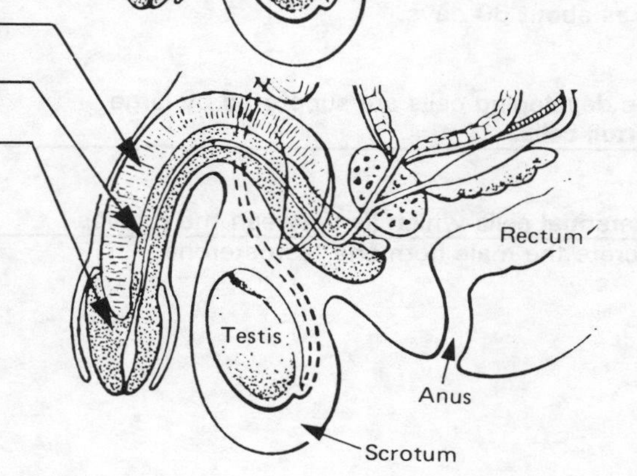

Erection of the penis is due to arterial dilatation controlled by the parasympathetic nervous system. The spongy spaces in the corpora become filled with blood at full arterial pressure.

Ejaculation is a reflex peristaltic contraction of the vas deferens and seminal vesicles which expels about 3 ml of semen containing about 300 million spermatozoa into the female genital tract during sexual intercourse.

1.2. Spermatozoa

The spermatozoa are produced in the seminiferous tubules. The lining of the seminiferous tubule is called the germinal epithelium.

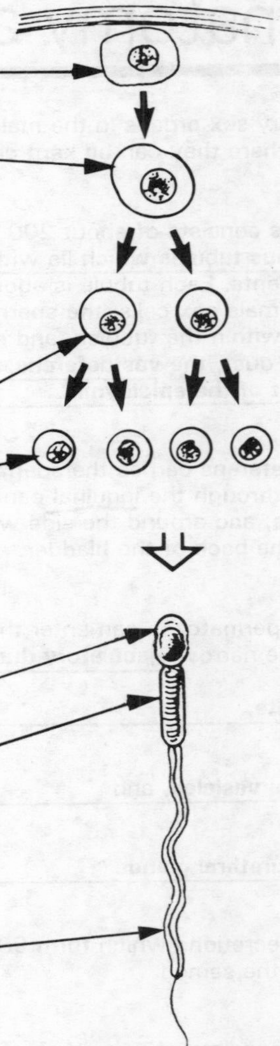

In the deepest layer, against the wall of the tubule, are cells known as the **spermatogonia.** In the adult these cells are constantly replicating by mitotic division. Some cells move inward and become **primary spermatocytes.**

These then undergo meiosis, in which the double (diploid) set of chromosomes found in all body cells is reduced to a single set. Subsequent union of male and female sex cells restores the diploid state.

The earliest cells with a single set of chromosomes in the male are the **secondary spermatocytes** which develop into spermatids. The **spermatids** undergo a complex modification to become spermatozoa.

The head of the spermatozoa is almost entirely nuclear material, rich in DNA and containing all the **paternal genes.**

Mitochondria are packed behind the head. The mitochondria contain the metabolic enzymes used to obtain energy from the surrounding fluid.

The **tail** is used for propulsion.

The development of spermatozoa in the tubules takes about 60 days.

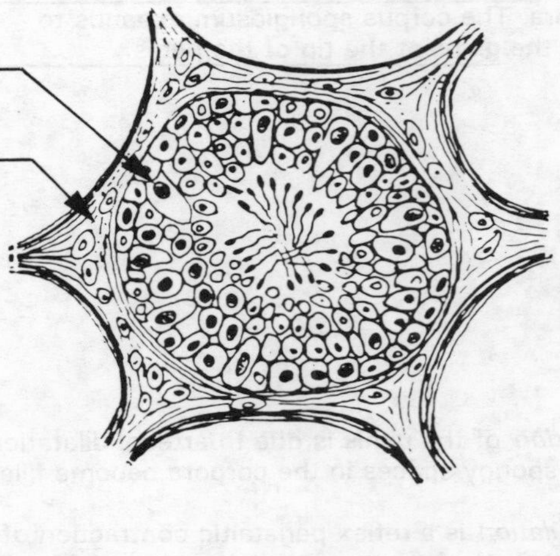

The developing cells are supported by large **Sertoli cells.**

Interstitial cells which lie between the tubules secrete the male hormone testosterone.

1.3. The male sex hormones

Between the ages of 10 and 15 years the anterior **pituitary gland** begins to secrete gonadotrophic hormones. This time of life is known as *puberty*.

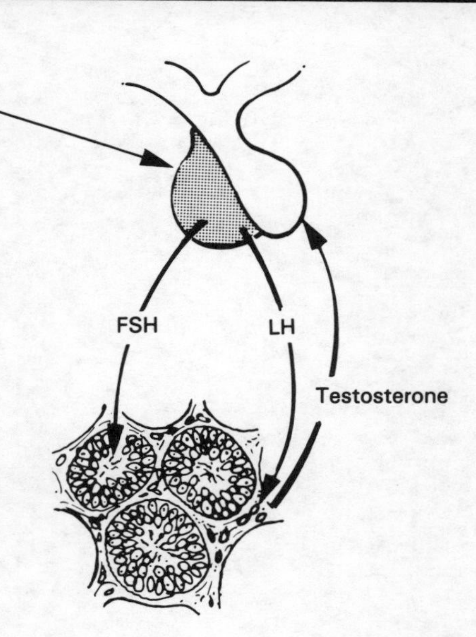

FSH (follicle stimulating hormone) initiates and maintains spermatogenesis by the tubules.

LH (luteinising hormone), also known as ICSH (interstitial cell stimulating hormone), causes secretion of testosterone by the interstitial cells.

The rate of secretion of LH (and partly of FSH) is controlled by inhibition of the pituitary by high levels of *testosterone*.

Testosterone is responsible for:

(i) growth, maintenance and secretion of the male reproductive tract, including the enlargement at puberty of the penis, scrotum and prostate;

(ii) activity of the germinal epithelium of the tubules (acting with FSH);

(iii) increased protein synthesis (this is the *anabolic* effect of testosterone responsible for a 'growth spurt' of bones and muscles at puberty);

(iv) growth of pubic hair, later growth of beard, axillary and chest hair, with development of frontal baldness;

(v) growth of the larynx, with deepening of the voice.

Testosterone secretion and spermatogenesis continue into old age with little decrease. There is no male equivalent to the female menopause.

TEST ONE

1. **What is the composition of semen?**

____

____

2. **Place the following in order to describe the pathway followed by spermatozoa.**

   (a) Urethra.
   (b) Epididymis.
   (c) Seminal vesicles.
   (d) Seminiferous tubules.
   (e) Vas deferens.
   (f) Ejaculatory duct.

3. **For which of the following is testosterone responsible?**

   (a) Inhibition of growth.

   (b) Increase of spermatogenesis.

   (c) Increase of the possibility of baldness.

   (d) Inhibition of the production of luteinising hormone.

ANSWERS TO TEST ONE

1. Semen consists of the spermatoza and the secretions of the prostate glands, the seminal vesicles and the bulbo-urethral glands.

2. Place the following in order to describe the pathway followed by spermatazoa.

(i) (d) Seminiferous tubules.
(ii) (b) Epididymis.
(iii) (e) Vas deferens.
(iv) (c) Seminal vesicles.
(v) (f) Ejaculatory duct.
(vi) (a) Urethra.

3. (b) Increase of spermatogenesis.
(c) Increase of the possibility of baldness.
(d) Inhibition of the production of luteinising hormone.

2. The female reproductive system

2.1. Anatomy and physiology

The primary sex organs in the female are the **ovaries**.

These secrete the hormones progesterone and oestrogen and produce the female sex cells, the **ova**.

Once a month, one ovum is produced. After its release from the ovary it is actively guided by movements of the **fimbriae** into the **uterine tube** which lies folded on the side of the uterus. It may be fertilised by a spermatozoon in the uterine tube.

The ovum then travels to the **uterus**, or womb. The uterus has a thick wall of smooth muscle and a vascular lining, *the endometrium*, which undergoes cyclical changes each month to prepare it for the acceptance of a fertilised ovum.

Normally the **uterine cavity** is flattened but it enlarges enormously during pregnancy to accommodate the fetus as it grows and develops.

The neck of the womb, or **cervix,** is at the entrance to the uterine cavity. It remains tightly closed, except during childbirth, although the spermatozoa can pass through.

The vagina is a fibromuscular tube usually flattened between the bladder and rectum. It is about 8 cm long and runs downwards and forwards to open behind the **pubic bone**. It sheaths the penis during sexual intercourse.

The **urethra,** the opening of the bladder, is a much smaller tube which lies in front of the vagina.

In front of the urethra is the **clitoris,** a sensitive mass of erectile tissue like a tiny copy of the male penis.

The **labia** are two folds of skin on the sides of these structures which hide them.

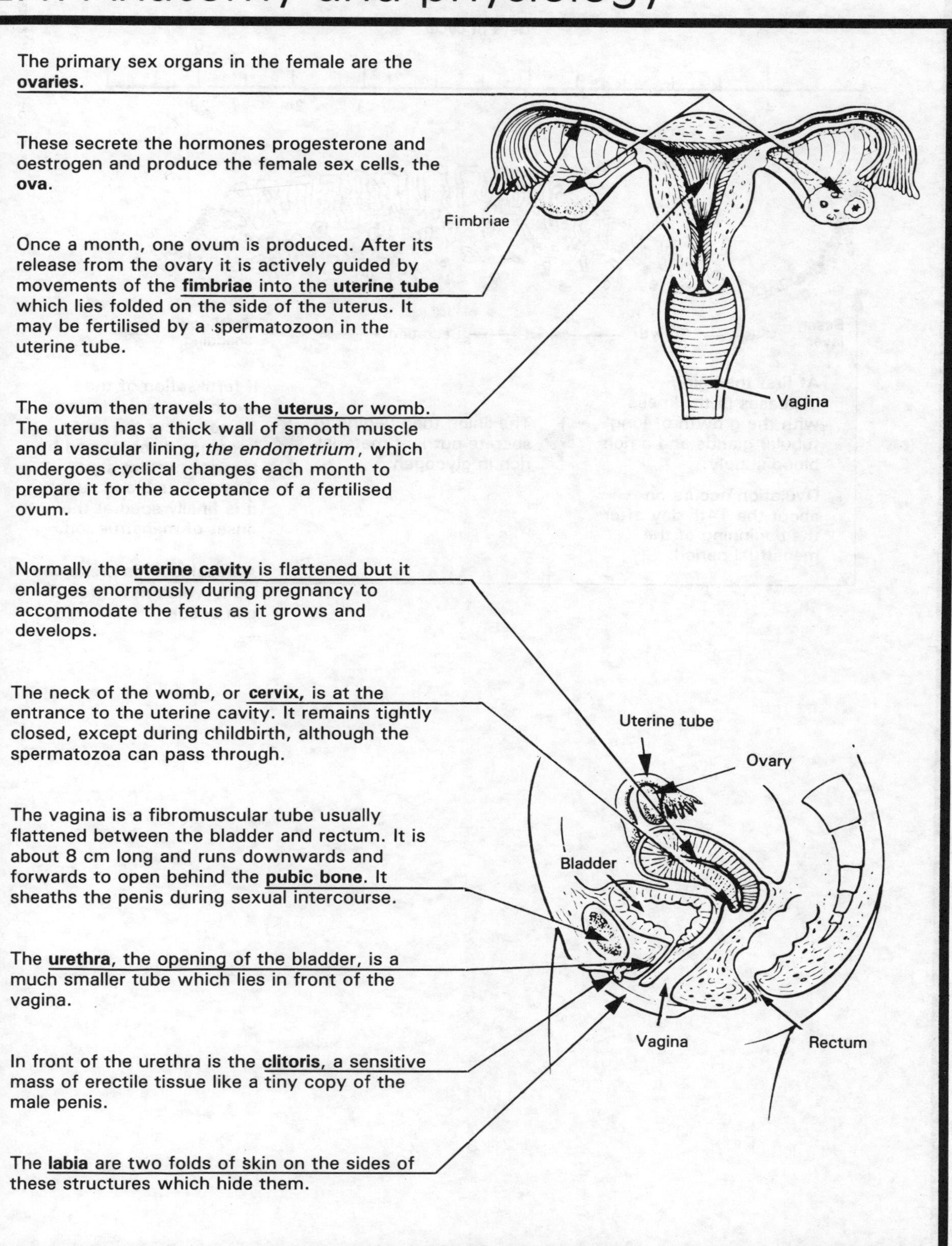

2.2. The menstrual cycle

The endometrium undergoes regular cyclical changes, each cycle lasting about 28 days and being marked by a loss of blood from the uterus. The changes are under the control of the hormones of the ovary, and ensure that the endometrium is always ready to receive a fertilised ovum on about the 21st day of the cycle.

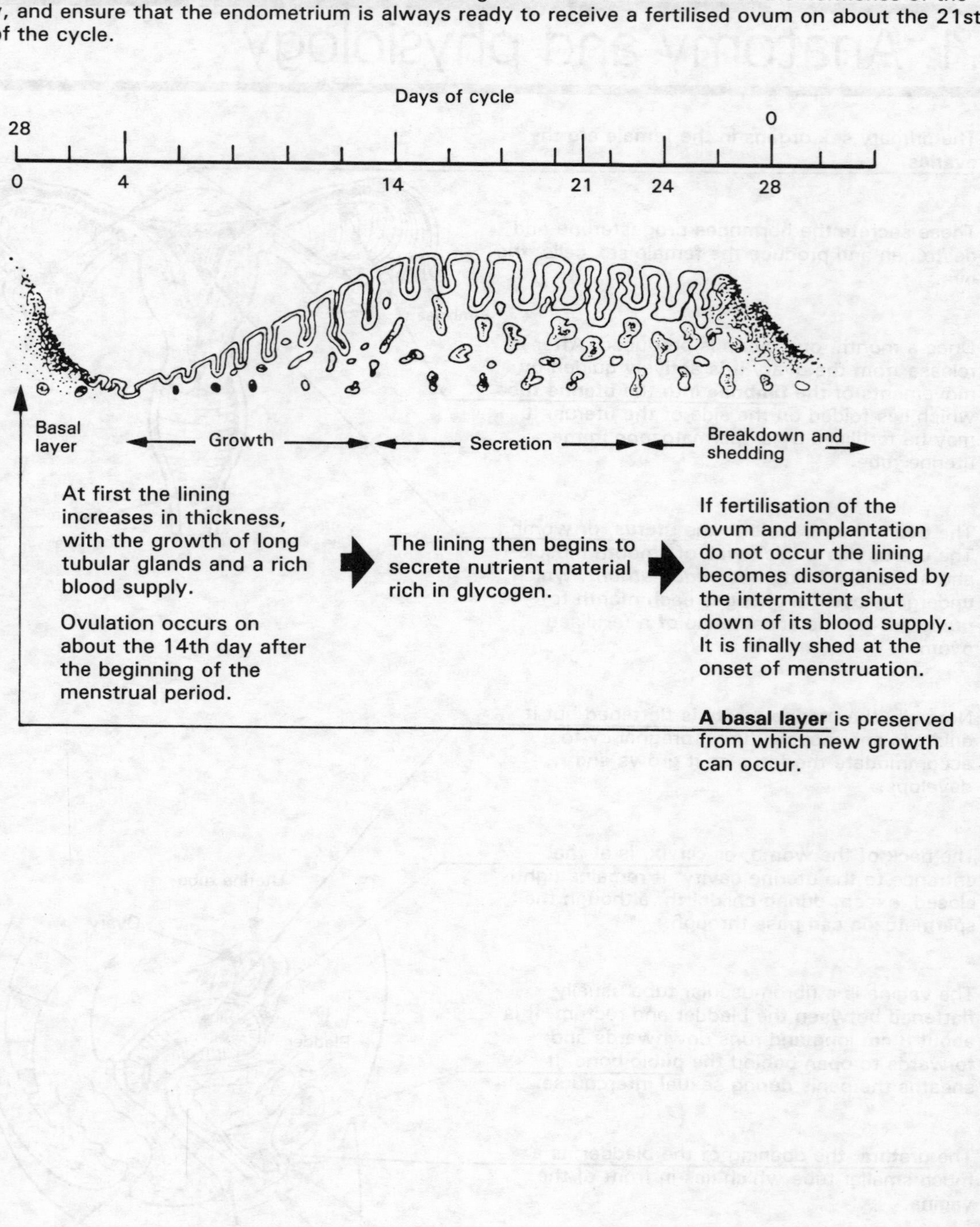

2.3. Ovulation

Each ovary contains about 18 000 ova in its substance. Each ovum is surrounded by a cluster of small cells, forming a **primordial follicle.**

All the primordial follicles have developed in the ovary by birth. No new ones form later.

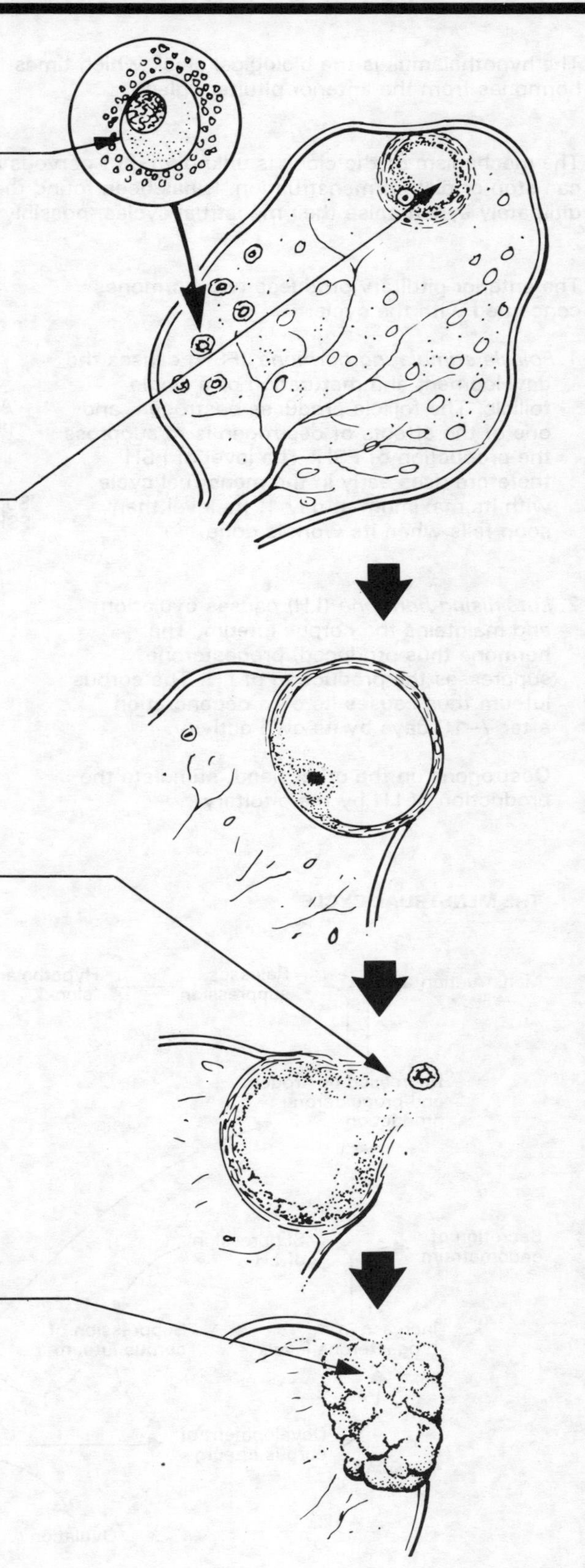

Each month a single follicle matures to form a **Graafian** (ovarian) **follicle**, a mass of cells surrounding fluid, with the ovum in its wall.

The Graafian follicle actively produces oestrogens. It enlarges until it is about 1 cm in diameter, and then bursts to shed the ovum, surrounded by a cloud of cells, into the **abdominal cavity.** The ovum is taken up by the fimbriae of the uterine tube.

The follicle collapses and its lining cells multiply to produce a solid yellow mass with a rich blood supply, the **corpus luteum.** This secretes the hormones progesterone and oestrogen into the blood for the next 7 to 10 days and then, if fertilisation of the ovum has not occurred, it atrophies to leave only a scar.

Progesterone acting with oestrogen produces secretion of the endometrium. Oestrogen alone is responsible for growth of the endometrium.

When both hormones cease being secreted the endometrium breaks down.

2.4. Hormonal control of menstrual cycle

The hypothalamus is the biological clock which times the menstrual cycle. It controls the release of hormones from the anterior pituitary gland.

The mechanism of the clock is unknown, but nervous influences can alter it. Anxiety or depression can stop or disturb menstruation. It has been found that groups of women living closely together ultimately synchronise their menstrual cycles, possibly through reflexes triggered by olfactory stimuli.

The anterior pituitary produces two hormones concerned with the cycle:

1. *Follicle stimulating hormone* (FSH) causes the development and maturation of a single follicle. The follicle produces oestrogen, and one of the effects of oestrogen is to suppress the production of FSH. The level of FSH therefore rises early in the menstrual cycle with its maximum at day 1. Its level then soon falls when its work is done.

2. *Luteinising hormone* (LH) causes ovulation and maintains the corpus luteum. The hormone thus produced, progesterone, suppresses the production of LH. The corpus luteum thus causes its own degeneration after 7–10 days by its own activity.

   Oestrogens on the other hand, stimulate the production of LH by the pituitary.

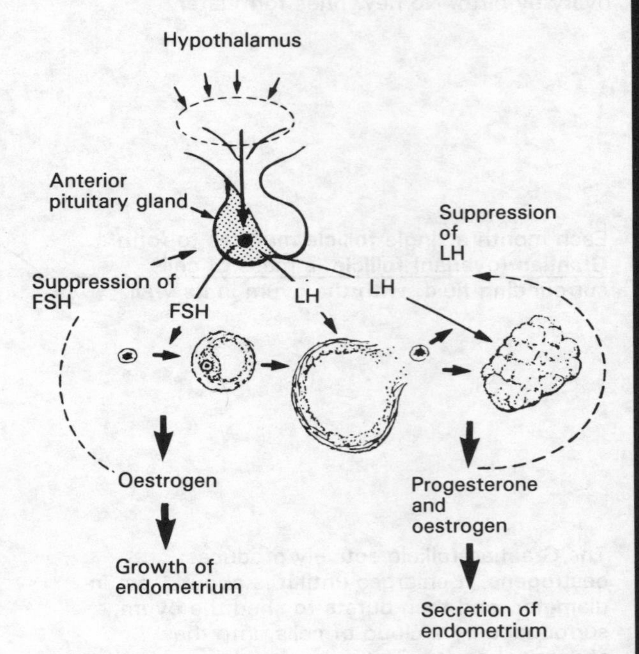

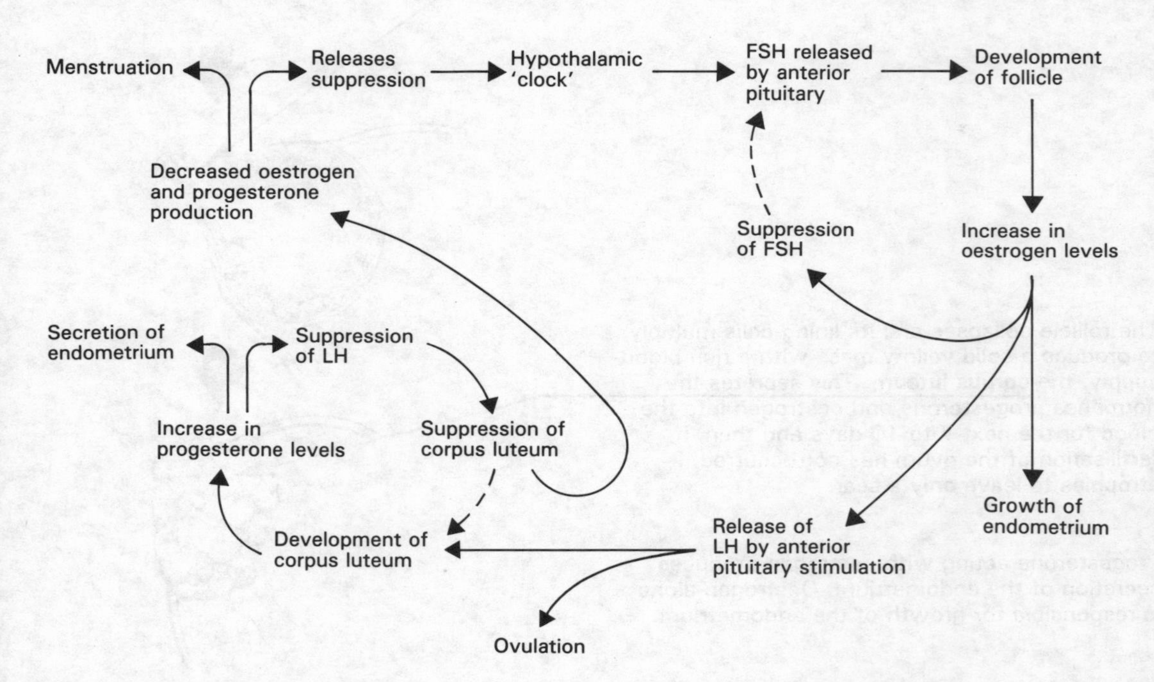

The effect of oestrogen and progesterone in suppressing the pituitary is made use of in the contraceptive 'pill'.

2.5. Puberty

The secretion of FSH and LH by the anterior pituitary begins between the ages of 9 and 12 years. Before this time the hypothalamus restrains the functioning of the anterior pituitary.

The hormones cause the ovary to produce oestrogens, but without the development of follicles at first. The ovaries enlarge.

Oestrogen causes:

(i) growth of the uterus and uterine tubes;

(ii) thickening of the vaginal epithelium;

(iii) development of the breasts;

(iv) development of pubic hair and axillary hair;

(v) development of typical female contours, with widening of the pelvis and a characteristic fat deposition on the thighs and lower abdomen.

The *menarche*, the onset of cyclical bleeding, tends to follow these changes. For the first few cycles follicles develop and then regress without ovulation. During these *anovular cycles* the endometrium does not secrete, but its thickened surface breaks down in the normal way. Such anovular cycles also occasionally occur in normal fertile adults.

2.6. The menopause

At some time between the ages of 40 and 55 years the ovaries stop producing ova and hormones. This is due to ovarian failure and results in an increase in the secretion of FSH and LH.

The decrease in circulating oestrogen and progesterone causes:

(i) cessation of menstruation — the *menopause*;

(ii) a decrease in the size of the uterus, the uterine tubes, the breasts and of the fat deposits on the thighs and lower abdomen;

(iii) a decrease in pubic hair and axillary hair.

The menopause may also be accompanied by vasomotor disturbances (causing 'hot flushes', blushing and sweating of the upper half of the body) and by emotional disturbances. Sexual activity and libido do not usually decrease, and may even increase.

TEST TWO

1. Indicate which of the names in the list below refer to the parts of the female reproductive system labelled on the diagram alongside, by placing the appropriate letters in the brackets.

1. Cervix ( )
2. Fimbriae. ( )
3. Corpus luteum. ( )
4. Endometrium. ( )
5. Uterine tube. ( )

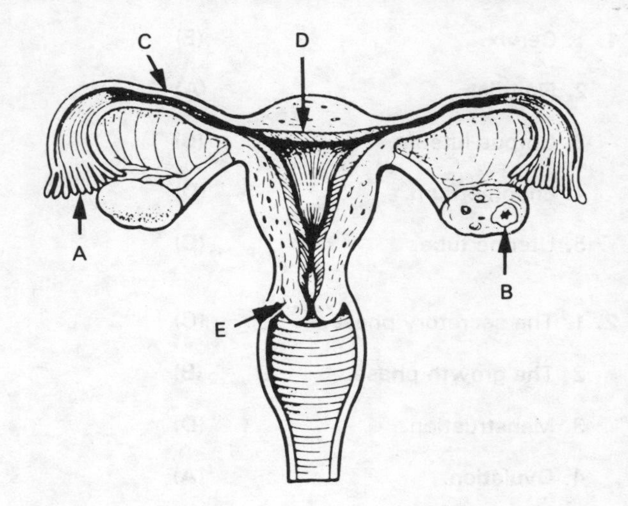

2. Indicate which of the names in the list below apply to the stages of the menstrual cycle labelled on the diagram alongside, by placing the appropriate letters in the brackets.

1. The secretory phase. ( )
2. The growth phase. ( )
3. Menstruation. ( )
4. Ovulation. ( )

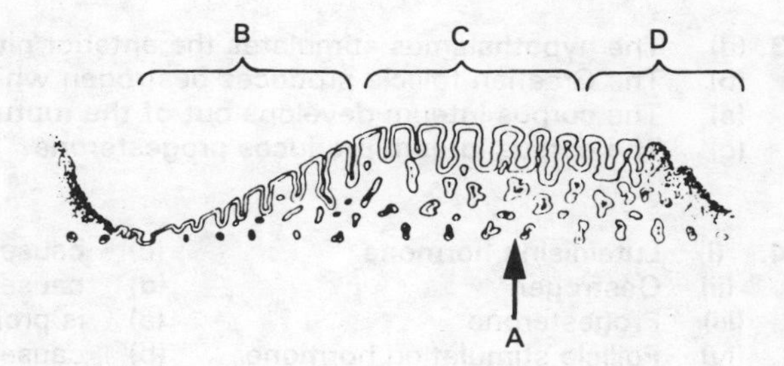

3. Arrange the following processes in the order of their occurrence:

(a) The corpus luteum develops out of the ruptured Graafian follicle.
(b) The Graafian follicle produces oestrogen while maturing.
(c) The corpus luteum produces progesterone.
(d) The hypothalamus stimulates the anterior pituitary gland to produce FSH and LH.

4. Which of the statements on the right apply to the hormones listed on the left?

(i)	Luteinising hormone.	(a)	Produced by the corpus luteum.
(ii)	Oestrogen.	(b)	Causes development of a follicle.
(iii)	Progesterone.	(c)	Causes ovulation.
(iv)	Follicle stimulating hormone.	(d)	Causes growth of the endometrium.

ANSWERS TO TEST TWO

1. 1. Cervix. (E)

2. Fimbriae. (A)

3. Corpus luteum. (B)

4. Endometrium. (D)

5. Uterine tube. (C)

2. 1. The secretory phase. (C)

2. The growth phase. (B)

3. Menstruation. (D)

4. Ovulation. (A)

3. (d) The hypothalamus stimulates the anterior pituitary gland to produce FSH and LH.
(b) The Graafian follicle produces oestrogen while maturing.
(a) The corpus luteum develops out of the ruptured Graafian follicle.
(c) The corpus luteum produces progesterone.

4. (i) Luteinising hormone — (c) causes ovulation.
(ii) Oestrogen — (d) causes growth of the endometrium.
(iii) Progesterone — (a) is produced by the corpus luteum.
(iv) Follicle stimulating hormone — (b) causes development of a follicle.

3. Pregnancy

3.1. Fertilisation and implantation

About 300 million spermatozoa are ejaculated by the male into the female genital tract. About 1 million of these swim through the cervix, where the mucus is thinned by oestrogenic activity during ovulation. A few hours later a few hundred spermatozoa reach the uterine tube. The spermatozoa have a useful life span of about 48 hours.

The ovum comes to lie in the uterine tube within an hour or so of ovulation. It remains able to be fertilised for about 48 hours.

The ovum is surrounded by a corona of small cells and by a thin membrane, the **zona pellucida**. The zona allows one, and only one, spermatozoon to penetrate it. This fuses with the ovum and its nuclear material swells. It joins with the female nucleus to restore the usual double chromosome complement. This set of chromosomes contains all the information for the production of a new individual, combining the maternal and paternal characteristics in a unique way.

The fertilised egg divides within the zona to form two, four, eight, sixteen and eventually a ball of cells with a central fluid-filled cavity. This is the **blastocyst.**

The blastocyst is gradually carried along the uterine tube, by peristalsis, to the uterine cavity where 5–7 days later it implants, usually in the upper uterus.

The zona prevents premature implantation of the blastocyst into the uterine tube, and then soon disintegrates.

The implanted blastocyst digests the endometrium it lies on and sinks into the thick, fleshy tissue to become completely buried.

The outer layer of the blastocyst, the **chorion,** invades the endometrium with finger-like projections called **villi.**

It absorbs nutrients and secretes the hormone human choriogonadotrophin (HCG).

HCG is a powerful hormone which imitates LH, maintaining the corpus luteum and preventing menstruation. (It can be readily detected in blood and urine and forms the basis of pregnancy testing.)

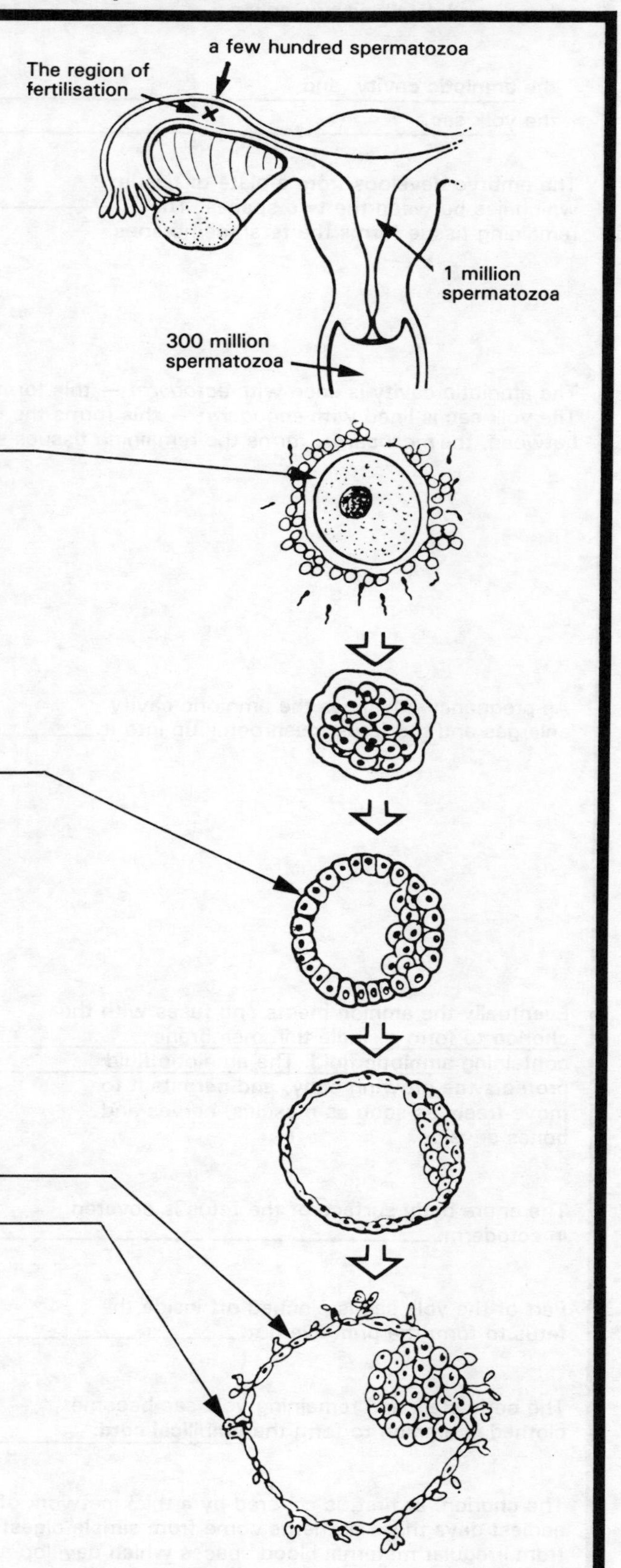

3.2. The development of the fetus

A thickening on one side of the blastocyst enlarges and develops two spaces:

the **amniotic cavity**, and

the **yolk sac**.

The embryo develops from a plate of tissue which lies between the two spaces. The remaining tissue forms the fetal membranes.

The amniotic cavity is lined with *ectoderm* — this forms the skin and nervous system of the fetus. The yolk sac is lined with *endoderm* — this forms the gut lining and viscera of the fetus. The tissue between, the *mesoderm*, forms the remaining tissues such as bones and muscles.

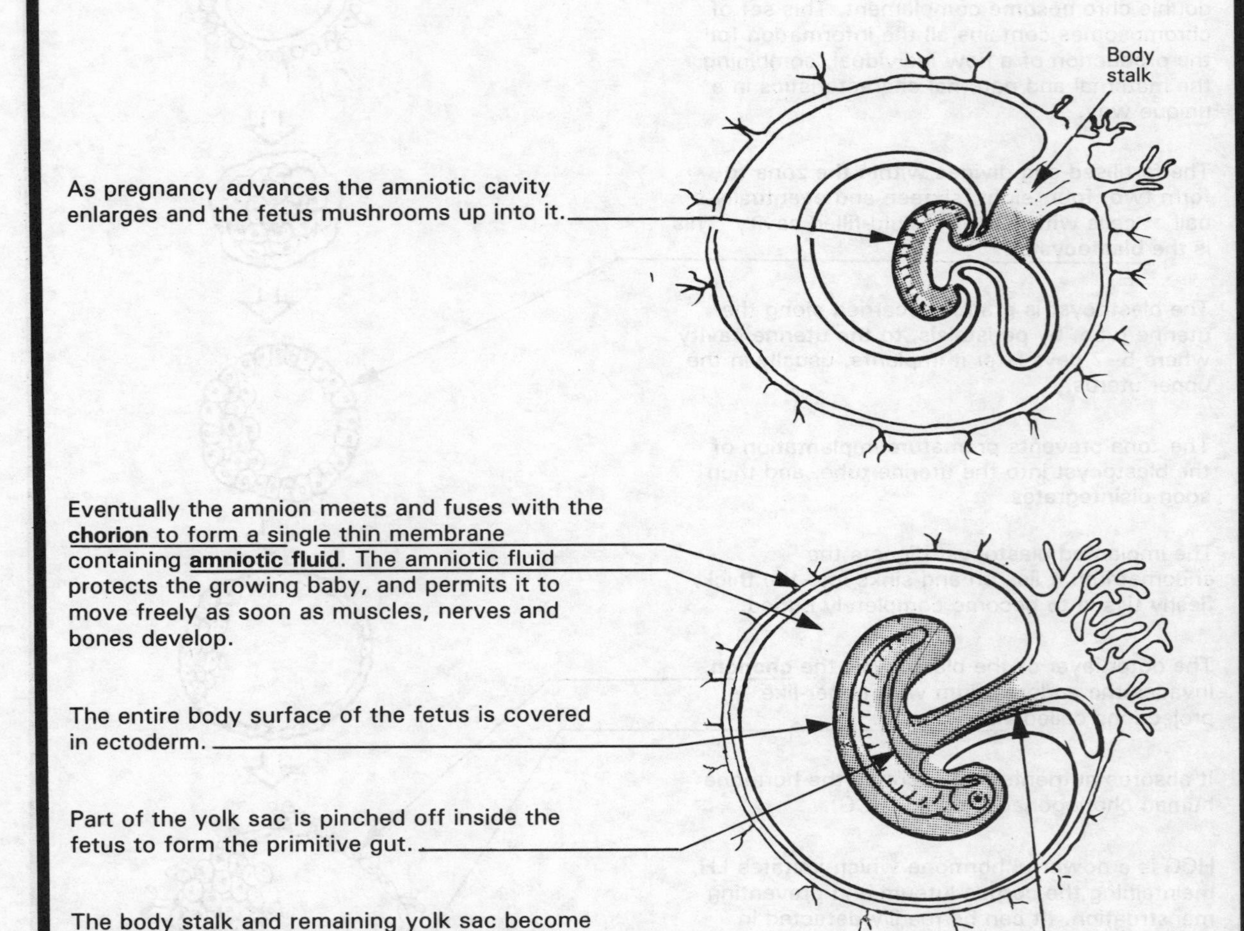

As pregnancy advances the amniotic cavity enlarges and the fetus mushrooms up into it.

Eventually the amnion meets and fuses with the **chorion** to form a single thin membrane containing **amniotic fluid**. The amniotic fluid protects the growing baby, and permits it to move freely as soon as muscles, nerves and bones develop.

The entire body surface of the fetus is covered in ectoderm.

Part of the yolk sac is pinched off inside the fetus to form the primitive gut.

The body stalk and remaining yolk sac become clothed in amnion to form the **umbilical cord**.

The chorion, at first, is covered by a thick network of branching villi, which absorb nutrients. In the earliest days these nutrients come from simple digestion of endometrial cells. Later they are absorbed from irregular maternal blood spaces which develop around the chorion.

With the formation of the umbilical cord, the area at its base becomes thickened and specialised as the **placenta**, a disc of tissue firmly attached to the uterine wall.

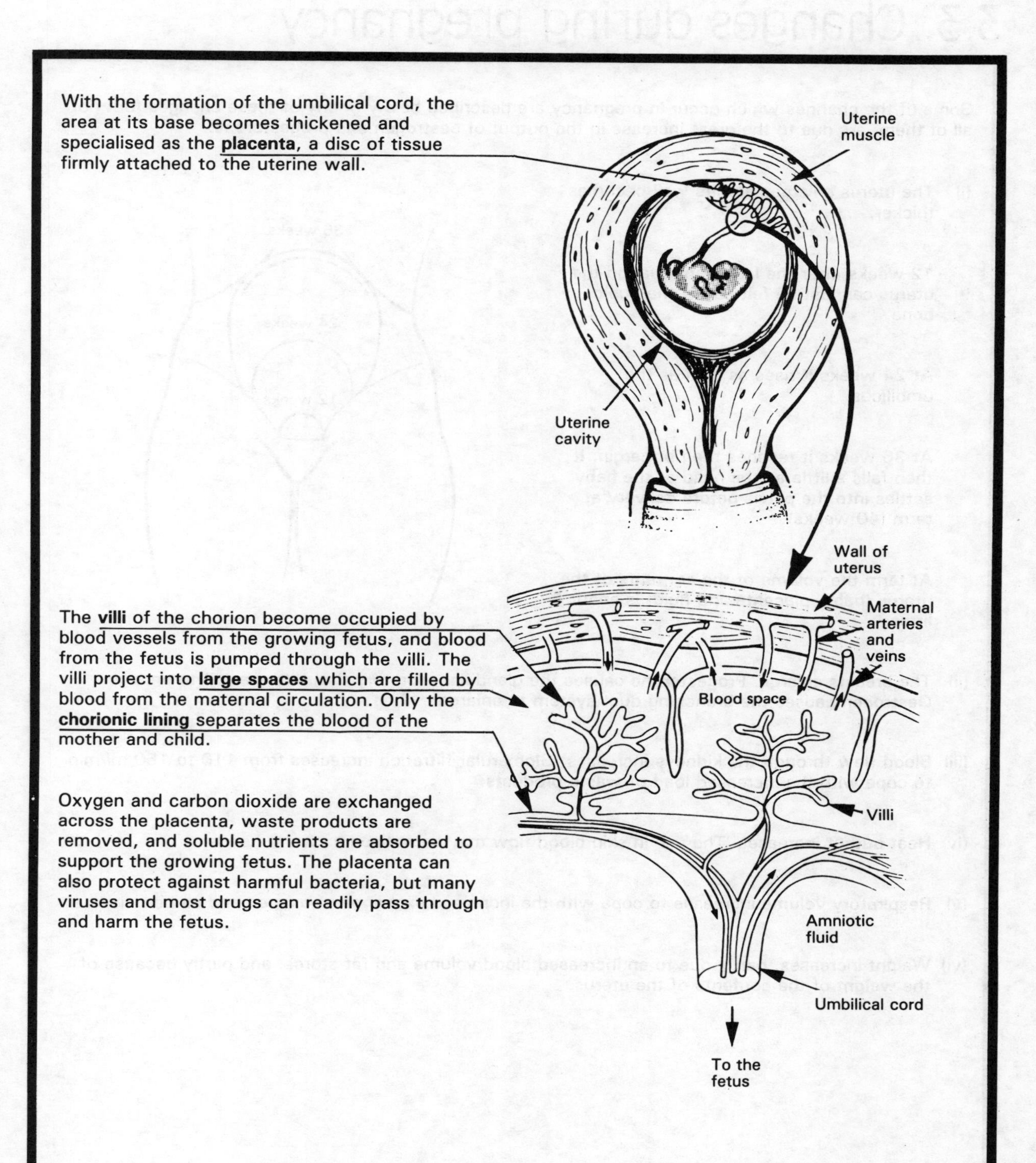

The **villi** of the chorion become occupied by blood vessels from the growing fetus, and blood from the fetus is pumped through the villi. The villi project into **large spaces** which are filled by blood from the maternal circulation. Only the **chorionic lining** separates the blood of the mother and child.

Oxygen and carbon dioxide are exchanged across the placenta, waste products are removed, and soluble nutrients are absorbed to support the growing fetus. The placenta can also protect against harmful bacteria, but many viruses and most drugs can readily pass through and harm the fetus.

The entire placenta, when fully developed, is about 18 cm across and 3 cm thick. It weighs about 500 g. It resembles a dark red sponge since the villi (which have a total surface area of about 10 sq metres) are tightly packed. Towards the end of pregnancy the placenta takes about 10% of the output of the mother's heart.

After 3–4 months the corpus luteum degenerates and production of progesterone and oestrogen is taken over by the suprarenal glands of the fetus and the placenta working together. The production of progesterone and oestrogen increases steadily throughout pregnancy.

3.3. Changes during pregnancy

Some of the changes which occur in pregnancy are described below. Many of these changes (but not all of them) are due to the great increase in the output of oestrogen and progesterone.

(i) The uterus enlarges and its wall becomes thicker.

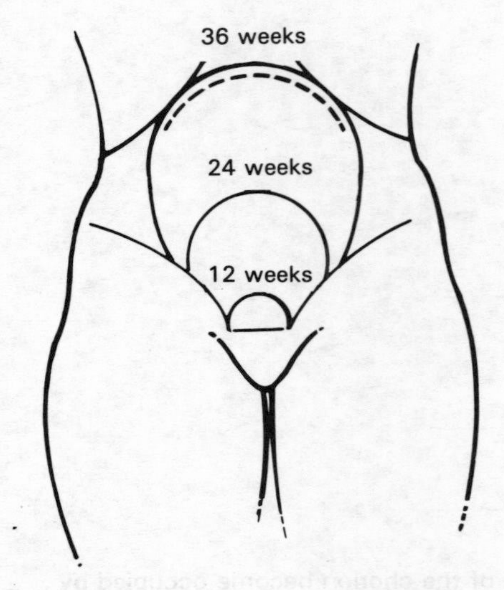

12 weeks after the last menstruation the uterus can just be felt above the pubic bone.

At 24 weeks it rises as high as the umbilicus.

At 36 weeks it reaches the rib margin. It then falls a little as the head of the baby settles into the pelvis before delivery at term (40 weeks).

At term the volume of the contents of the uterus (baby, placenta and fluid) is about 5 litres.

(ii) The breasts enlarge. Progesterone causes the glandular secretory tissue (the acini) to enlarge. Oestrogen causes the branching duct system to enlarge.

(iii) Blood flow through the kidneys increases. Glomerular filtration increases from 110 to 150 ml/min to cope with the increased load of waste products.

(iv) Heat output increases. The rise in skin blood flow may be noticed as warm hands.

(v) Respiratory volume increases to cope with the increased exchange of oxygen and carbon dioxide.

(vi) Weight increases, partly due to an increased blood volume and fat stores, and partly because of the weight of the contents of the uterus.

3.4. Birth

Birth, or *parturition*, occurs about 40 weeks after the last menstruation and 38 weeks after the fertilisation of the ovum.

In the few weeks before birth the head of the baby enters the pelvis and the cervix becomes notably softer in consistency.

At term the uterine muscle begins to contract in slow waves. At first this pulls the cervix over the baby's head and dilates it. The contractions may last for several hours and at some point the membranes rupture and the amniotic fluid is released.

When the cervix is fully dilated the uterus increases its contractions and rapidly expels the baby.

A few minutes later the contractions restart. The placenta separates completely and is shed with the membranes.

The uterus returns to its previous size over the next few days.

The factors which cause labour to start are not known, but once it it established the hormone oxytocin is released from the posterior pituitary to reinforce and maintain uterine contraction. Oxytocin is released reflexly in response to stretching of the cervix.

3.5. Lactation

Each breast consists of about 20 branching **ducts** which open via a **sinus** onto the surface of the nipple. There are supporting strands of fibrous tissue attached to the chest wall, and there are many fat cells between the lobules.

The duct system is well established after puberty, due to the influence of oestrogen, but the secretory **acini** only develop in pregnancy under the influence of high levels of progesterone. *Prolactin*, a hormone from the anterior pituitary, increases the action of both oestrogen and progesterone.

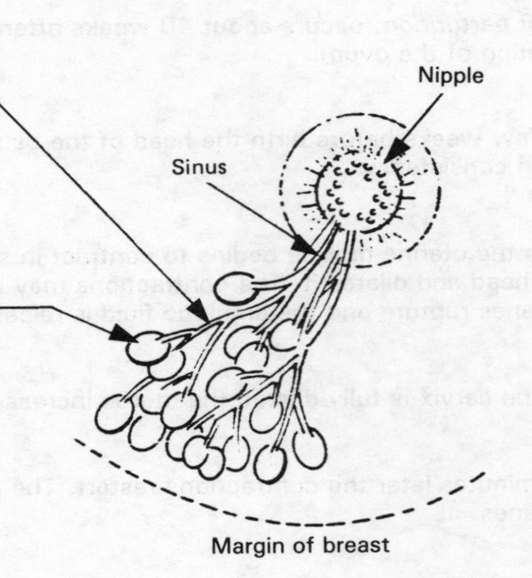

After childbirth falling oestrogen and progesterone levels cause an increase in prolactin secretion, and this stimulates the secretion of milk by the acini. The first secretion is *colostrum* a protein-rich fluid which contains antibodies. After the third day normal lactation is established.

Suckling of the baby at the breast stimulates the nipple which causes reflex secretion of the hormone *oxytocin* from the posterior pituitary. Oxytocin causes contraction of the smooth muscle fibres around the acini and milk is rapidly ejected from the nipple. A reflex, known as the 'let down', is established in the first few days of breast feeding but is readily disturbed by emotion. The release of oxytocin also helps the uterus to contract to its normal size.

TEST THREE

1. Why would it be true to say that the zona pellucida functions solely as a control?

____

____

2. Indicate which of the names in the list below apply to the features labelled on the diagram alongside, by placing the appropriate letters in the brackets.

1. The amniotic cavity. ( )
2. The yolk sac. ( )
3. The fetus. ( )
4. The chorion. ( )
5. The villi. ( )
6. The body stalk. ( )

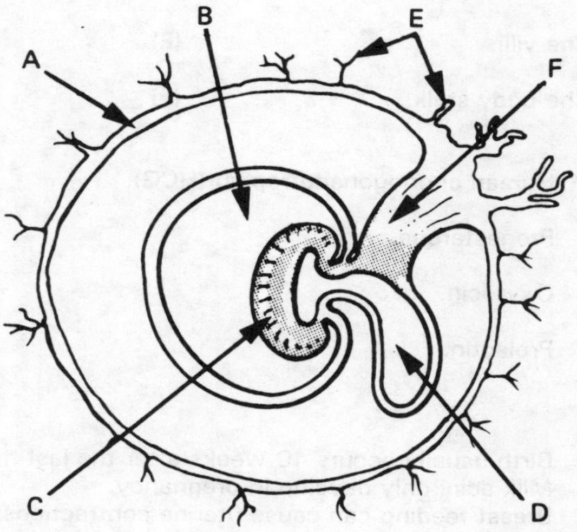

3. (a) Which hormone maintains the corpus luteum during the early stages of pregnancy?

(b) Which hormone causes the development of secretory acini in the breasts during pregnancy?

(c) Which hormone is released from the posterior pituitary to reinforce and maintain uterine contraction?

(d) Which hormone stimulates the secretion of milk by the acini?

4. Are the following statements true or false?

		True	False
(a)	Birth usually occurs 40 weeks after the last menstruation.	( )	( )
(b)	Milk acini only develop in pregnancy.	( )	( )
(c)	Breast feeding can cause uterine contractions.	( )	( )
(d)	The placenta weighs about 500 g.	( )	( )

ANSWERS TO TEST THREE

1. The zona allows only one spermatozoon to penetrate into the ovum and then later, having prevented premature implantation of the blastocyst into the uterine tube, it disintegrates.

2. 1. The amniotic cavity. (B)

2. The yolk sac. (D)

3. The fetus. (C)

4. The chorion. (A)

5. The villi. (E)

6. The body stalk. (F)

3. (a) Human choriogonadotrophin (HCG).

(b) Progesterone.

(c) Oxytocin.

(d) Prolactin.

4.

		True	False
(a)	Birth usually occurs 40 weeks after the last menstruation.	(√)	( )
(b)	Milk acini only develop in pregnancy.	(√)	( )
(c)	Breast feeding can cause uterine contractions.	(√)	( )
(d)	The placenta weighs about 500 g.	(√)	( )

POST TEST

1. Indicate which of the names in the list below apply to the parts of the male reproductive system labelled on the diagram alongside, by placing the appropriate letters in the brackets.

1. The prostate. ( )
2. The urethra. ( )
3. The corpus cavernosa. ( )
4. The vas deferens. ( )
5. The seminal vesicles. ( )

2. Indicate which of the names in the list below apply to the parts of the female reproductive system labelled on the diagram alongside, by placing the appropriate letters in the brackets.

1. The uterus. ( )
2. The vagina. ( )
3. The urethra. ( )
4. The clitoris. ( )
5. The labia. ( )

3. What is the function of the endometrium?

____

____

____

4. Which of the following contain a single set of chromosomes?

(a) A spermatid.

(b) A fertilised ovum.

(c) An ordinary body cell.

(d) An ovum just after ovulation.

POST TEST

5. (a) What stage of development is shown by the diagram alongside?

(b) What are the projections on its surface called?

(c) Why does it secrete the hormone HCG?

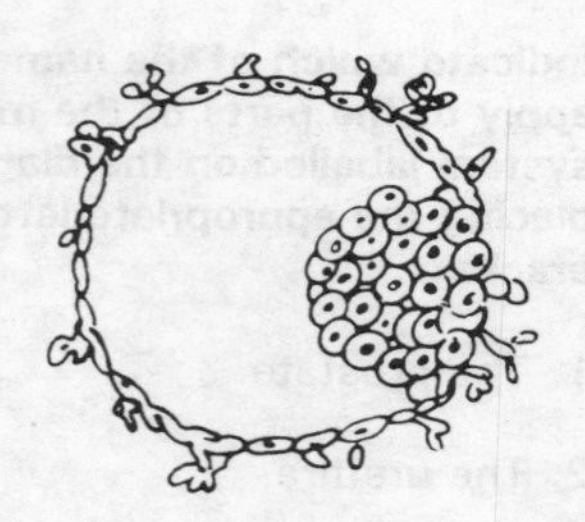

6. Which of the statements on the right apply to the items listed on the left?

(i)	The yolk sac.	(a)	Comes to surround the external surface of the fetus.
(ii)	The amniotic cavity.	(b)	Lies in contact with the endometrium.
(iii)	The ectoderm.	(c)	Partly forms the primitive gut.
(iv)	The mesoderm.	(d)	Later forms the nervous system.
(v)	The chorion.	(e)	Later forms bone and muscle.

7. Place the following events in the order in which they occur:

(a) The onset of secretion of FSH and LH by the pituitary.
(b) The onset of ovarian secretion of oestrogen.
(c) The onset of breast development.
(d) The onset of ovulation.
(e) The onset of cyclical uterine bleeding.

ANSWERS TO POST TEST

1. 1. The prostate. (E)

2. The urethra. (B)

3. The corpus cavernosa. (C)

4. The vas deferens. (A)

5. The seminal vesicles. (D)

2. 1. The uterus. (E)

2. The vagina. (D)

3. The urethra. (C)

4. The clitoris. (A)

5. The labia. (B)

3. The endometrium, the uterus lining, is the reception site for the fertilised ovum. The changes undergone by the endometrium in the course of the menstrual cycle ensure its readiness for receiving the ovum.

4. (a) A spermatid.

(b) An ovum just after ovulation.

ANSWERS TO POST TEST

5. (a) A blastocyst.

(b) Villi.

(c) To prevent menstruation.

6. (i) The yolk sac — (c) partly forms the primitive gut.
(ii) The amniotic cavity — (a) comes to surround the external surface of the fetus.
(iii) The ectoderm — (d) later forms the nervous system.
(iv) The mesoderm — (e) later forms bone and muscle.
(v) The chorion — (b) lies in contact with the endometrium.

7. (i) (a) The onset of secretion of FSH and LH by the pituitary.
(ii) (b) The onset of ovarian secretion of oestrogen.
(iii) (c) The onset of breast development.
(iv) (e) The onset of cyclical uterine bleeding.
(v) (d) The onset of ovulation.